In Deep
The Story of Global Diving & Salvage

by Joe Gustav

Documentary Media LLC, Seattle, Washington

In Deep: The Story of Global Diving & Salvage

First Edition 2014
Printed in Canada

Author: Joe Gustav
Project Managers: Frank Immel, Jimena Castria
Book Designer: Marilyn Esguerra, Philips Publishing
Editor: Judy Gouldthorpe
Editorial Director: Petyr Beck
Editorial Consultant: Robert Spector Consulting

ISBN: 978-1-933245-36-2

Library of Congress Cataloging-in-Publication Data

Gustav, Joe, 1988-
In deep : The Story of Global Diving & Salvage / Joe Gustav. – First edition.
pages cm
ISBN 978-1-933245-36-2
1. Global Diving & Salvage, Inc.–History. 2. Deep diving–United States–History.
3. Salvage–United States–History. 4. Underwater construction–United States–History.
5. Search and rescue operations–United States–History. 6. Marine accidents–United States–History.
7. Offshore oil industry–United States–History. I. Title.
VM981.G87 2014
338.7'627720973–dc23
2014002165

Published by:

Documentary Media
3250 41st Avenue SW, Seattle, WA 98116
(206) 935-9292 I books@docbooks.com
www.documentarymedia.com

ROBERT SPECTOR CONSULTING
Philips Publishing Group

Contents

dedications

This exercise would not be possible without first mentioning my true partner in this endeavor, Jeri. She has been the guiding light, the support mechanism, the setter and corrector of the moral compass, the leveler, and the plumb bob of the frenetic organization that is Global. She has kept me grounded (as much as possible) and helped me keep an eye on the prize and realize that the real target differs from day to day: one day it's family, one day it's friends, and one day it's Global. She has allowed me the latitude to balance all these important parts of my life and has been there to correct my course when needed.

This is a sea story, and the beauty of a good sea story is that it tells some details, adds some drama and laughter, but always makes a point. As you read our little sea story, know that there is a point to each section, and we hope that in the end you will see an ultimate sketch of what it is we've been trying to accomplish in these past decades, aside from making a living. Also remember that a wise old human once said, "The tale grew in the telling." Yes, it does grow and become a bit more dramatic and less like a textbook, but the essence should remain at the forefront.

I want to recognize all the folks who've been involved all these many years. The ones who've come and gone. The ones who've come and gone and come again. And all who've stuck with us, and bought into our little vision of what we think a work environment and product should be.

And of course, Tim. We've come a long way. Singing the same song, sometimes in different keys, but always with the same goals and same standards in mind. We've always shared a vision built on safety, integrity, and the highest possible quality in our work.

To all the rest of the employees, partners, clients, and associates, if you were not recognized here, you are recognized in this book.

Regards,

John Graham, cofounder, Global Diving & Salvage

This book tells one hell of a story, an unlikely tale when you get down to it. It is a story showcasing a remarkable company, a tale of luck and hard work, vision and confusion, grand successes and inevitable failures. Through it all, Global, our company, has been the focus. Not founder and co-owner John Graham or I, not any one diver, salvor, or beach foreman, just our little company that we had big dreams and aspirations for way back in the '80s. JG and I shared a secret dream and we have had a truly miraculous run as partners and friends. I challenge anyone to match what we have done together, and are we ever grateful—grateful to our crews, our leaders, the old-timers like Jackie, Dan Pierson, Dan Gilchrist, and Jeff Eyler, all of the team. But mostly I am grateful for John, a solid friend for over 40 years.

And while the opportunity is within grasp, let it go on record that my family and personal friends have put up with a lot, and still stand firm. Long jobs, endless travel, and many distracted moments have not stood in the way. Love and gratitude to all, especially my loving wife, Nerissa; the kids, Madeline, Melissa, Megan, Margo, and Miles; and yes, the grandkids too, Lucy, Stella, Mac, and sweet little Sally.

Jack Beaver, my pop, lent the money to buy Global's first dive boat, the M/V *Bottom,* away from the infamous "Bippo," and guided me to be an independent businessman. We looked upon my dad as an unofficial board member back in the day, and he was always full of advice and encouragement. Thanks, Pop.

Still in many ways the journey has just begun, and Global's history is being written every day. The legacy that is Global grows proudly and steadily. Onward, Global, onward—next stop, 50 Years!

Tim Beaver, Global Diving & Salvage

Tim Beaver's dad in front of the Salvage Chief *in Astoria, shortly after Global bought into Fred Devine Diving & Salvage Co. with Jim Dunlap and Mick Leitz.*

Foreword

Quite often, when a company reflects upon its history, there are a variety of perspectives and points of view that shape the story. Ours is no different. The history of our company is colorful and exciting; challenging and frustrating; wonderful and sad all at the same time. And at the end of the day, the story of Global is not about the projects, or the challenges, or the "little dive company just trying to scratch out a living." It is about our employees. Every single one of us who has had the privilege to work here is a critical thread in the overall tapestry of Global.

Throughout the book, there are many recollections and stories, some honest to God's truth, some made much clearer after a few drinks to loosen the memory, and some perfectly revisionist history by the storyteller. The projects and events detailed are but a few select examples of the hundreds of projects large and small that are accomplished on a daily, weekly, monthly, and yearly basis. Many individuals are mentioned by name, and many are not. But all are important to the legacy of Global. No one event, project, or individual is more important than the rest; each has contributed to the larger effort. The philosophy that has served us well for over 35 years remains as true today as it always has been: sincerity and honesty in our business, safety and competency in the workplace, and professional passion and pride in our craft.

So thank you to all those who have contributed to Global's success over the years, and to making tomorrow better by working hard today. And know that we are all contributors to the future unwritten chapters of the Global story.

Best,

Devon Grennan, CEO/President

chapter 1

Getting Wet

> **Divers: A Problem Since the Beginning**
>
> DIVER SAYING PRINTED ON A T-SHIRT FOR THE NOBLE ENERGY JOB, MID-2000s

Professional divers say that diving itself is just a mode of transportation. What really matters—what sets the pros apart from the weekend sport divers, and what makes or breaks individual careers and entire companies—is what a diver can do underwater. And what exactly a diver can do underwater, working alone in the cold and dark like an earthbound astronaut going where humans are not supposed to be able to go, is determined by three factors. First, skill across multiple disciplines, from welding to carpentry; second, the ingenuity of his support group on the surface; and third, the needs of the job at hand.

Of course, no two jobs are ever the same. Each presents its own set of challenges. Regardless of the circumstances surrounding the project, from salvage to ship husbandry, the diver's goal after "getting wet" is to solve the client's problem while maintaining the safest conditions possible. Often the greatest challenge dive crews face is educating the client as to what can actually be done and what is possible, as well as what is happening below the surface. This realignment of realities is the origin of a comment frequently heard from clients: "Well, these divers have been a problem since the beginning."

The tune changes quickly as the job moves toward an efficient, effective, and safe conclusion, the kind of result the diving crew from Global Diving & Salvage was struggling to achieve as the sun set on a gray Northwest evening in 1988. The Global crew had seen their fair share of catastrophe, from oil spills to shipwrecks, since forming

A Global diver on the salvage of the Deep Sea *in Penn Cove, Washington.*

in 1979. The task for Tim Beaver and his crew was to bring a vessel that had sunk to the ocean floor back to the surface. More specifically, their job at this very moment was to support Beaver's partner John Graham, who was diving beneath them, patching leaks in an effort to find the right one that would allow for the vessel to finally be raised. A dive job ultimately falls on the shoulders of the person working alone in the water. However, there's much more to a successful dive than a guy going underwater—a support team is there to first ensure that the diver can do his job safely, then to have the right plan and tools at hand to do high-quality work. This is to say nothing of off-site personnel who bid projects, win them, and then make the arrangements for the divers to go do their job.

It had been a long day. Light was dwindling and everyone was tired, especially Graham, who had spent his last four-hour rotation searching for any kind of leak that would make the difference so they could raise the vessel inch by inch using gas-powered pumps that had been staged earlier. Hours ago when they'd tried starting the pump, the vessel gave no ground. Now they would try again.

One pump started, adding to the cacophony of machinery and movement heard on a dive job site. Then another pump roared, now the two of them in unison before being joined by a chorus of two more. The air filled with black smoke and the smell of exhaust was pervasive. At last the vessel jolted upward, ever so slightly.

"A vessel raising on a salvage is one of the most exciting things you could ever be around," said Greg Harem, founding member of Global Diving and the company's first president. "There is a huge rush of accomplishment. It happens quickly. It's different than finishing a construction job. You don't know for sure when it's going to take off, but all of a sudden, it breaks loose and you're on your way." The crew looked on as the vessel rose out of the cold Puget Sound waters, overcome by the warm feeling of accomplishment experienced upon the successful completion of a complex job.

Top: John Graham surveys a sunken vessel.

Bottom: A tender assists a diver in preparation to enter the water.

Opposite: A diver installs an anode amidst thick barnacles.

A Dive Helmet History

THE YOUNG DIVERS WHO WOULD GO ON to form Global Diving & Salvage arrived on the scene at a time when dive helmets for surface-supplied dives were changing for the first time in over a century. The standard model for "hats," as divers call them, was invented in 1837 by Augustus Siebe, credited as the Father of Diving. His was the first "closed" helmet that could be bolted to a diving suit to create a watertight seal. The helmet was made of tinned copper to prevent corrosion caused by sea water, and so was known as the "tin hat."

The tin hat was the dominant model of helmet, changing little except for some advances in comfort and surface communication until the 1960s and the advent of first, compact stainless steel helmets by Savoie, and then the fiberglass helmet. These helmets also introduced demand regulators that delivered air to the diver as he or she breathed. Divers sought to blend the mobility of scuba technology with the safety and communication advantages of a hard hat, and so transitioned from the clunky tin hat to the relatively modern band masks made by Aquadyne and Kirby Morgan.

The pre-Global Diving crew at Crowley Maritime found themselves in the midst of this change and helped bring the technology to the Pacific Northwest. "We brought it from the North Sea, where John Graham and others dove," Tim Beaver said. "It was not a welcome technology for some divers in Puget Sound. We were derided to some degree for wearing the new helmets, which the old-timers nicknamed Tupperware. We couldn't have cared less—they were more efficient, safer, and used a hell of a lot less air."

The fiberglass helmet is still the prevalent model today. It has received some significant upgrades in terms of diver comfort, improved regulators, and much clearer and more consistent surface communication. Dive professionals continue to make small tweaks that allow for lighter-weight helmets, better visibility, and a better fit, all while improving safety.

Left to right: Painting of Augustus Siebe by George Beechey, circa 1840; a modern helmet used by professional divers.

The team of divers and support staff at Global Diving & Salvage have experienced the rush described by Harem time and again over their 35 years, a sense of accomplishment that is only heightened by the immensity of the business risks they have taken and the creative solutions they have devised to tackle increasingly complicated and sensitive projects. Global Diving now has four regional offices in the U.S. and works around the world, completing major construction projects, cleaning up oil spills, and salvaging vessels of all kinds, all with a fundamental dedication to the safety of the crews, the quality of their work, and the integrity of their methods. The company's roots are much humbler than their more recent large-scale projects around the world might suggest, beginning in the days of tin hats and many a paycheck spent at Harbor Island haunts along Seattle's waterfront.

From the Ancient Greeks to the Alki Tavern

The famed conqueror Alexander the Great supposedly said, "There is nothing impossible to him who will try." Alexander himself attempted to solve one of humankind's great limitations: our inability to breathe underwater. By several accounts, he had constructed and was himself lowered into the water in a crude diving bell made of white glass in the 4th century BC. But Alexander was not the world's first diver—the Chinese dove for oyster pearls as early as 2200 BC, and it's likely that military divers fought in the Trojan War.

Over the next 3,000 years, some of the world's greatest minds attempted to improve on primitive designs for underwater breathing apparatuses, including Leonardo da Vinci, who connected cane tubes to a leather face mask and tested his design in shallow water. It was not until the late 1830s that a closed dive helmet with air pumped from the surface to the diver was invented and became the prevalent mode of diving, changing relatively little over the next century and a half.

Sixteenth-century painting of Alexander the Great being lowered in a glass diving bell.

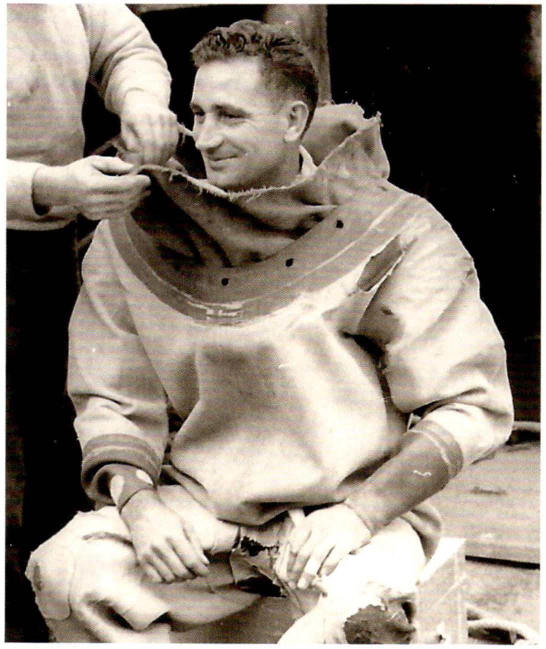

By the 1900s, diving was an established trade, helping expand coastal cities out into the water with lengthy piers and assisting in the maintenance of vessels that determined global power and facilitated the exchange of goods and immigration over great distances.

Where there were bustling ports and marine industry, there were divers, and the Pacific Northwest was no exception. Much of the shipyard action in Seattle took place on Harbor Island, which at the time of its completion in 1909 was the largest artificial island in the world. Todd Pacific Shipyard and Lockheed Shipyard, both on Harbor Island, were the dominant shipbuilding and repair yards in the Pacific Northwest, employing 5,000 people in their heyday during the war efforts of the 1940s.

Typically, divers who did ship husbandry and repairs were freelancers working in two-person teams comprised of a diver and a tender, usually a fellow diver who provided above-water support. When they got the call for work, they would show up with their large box of dive gear, the "hat" (or dive helmet) resting on the lid. The work was hard, and they lived hard lives, often joining their shipyard counterparts after work at one of Harbor Island's preferred watering holes. By the 1950s, around 11:50 p.m. the sound of squealing tires and flying gravel would signal the end of a shift at nearby Todd or Lockheed and the beginning of the migration to the Blew Eagle or Nifty's, the latter famous for its whiskey and pancakes. Years later in the 1970s, a little farther down the road overlooking the shipyards was the Alki Tavern, on West Seattle's Harbor Avenue. The Alki would become the preferred establishment for an ambitious younger generation of divers.

Top: Harbor Island in 1943.

Bottom: Ed Burroughs, grandfather of current Global employee Sarah Burroughs, was a diver for the U.S. Navy. In the late 1950s he participated in a radio promo in which his voice was broadcast from underwater for a station in North Bend, Oregon.

Diver Leiter Hockett and his crew building shipways in Grays Harbor, Washington, in 1948.

A Changing of the Guard

The freelance two-man teams eventually gave way to small dive companies when larger maritime contractors needed reliable, consistent dive work. Increased safety regulations that required dive plans, safety records, and specific insurance for workmen working as divers were a motivator. Another major impetus came with the passing of the National Environmental Policy Act in 1969, which among other requirements mandated that companies clean up maritime oil spills. There was now a need for oil spill response but no one to fill the demand, and so several companies including Seattle-based Crowley Maritime formed Marine Oil Pickup Services, or MOPS. When Crowley's partners eventually backed out, the operation was renamed Crowley Environmental Services (CES). In addition to environmental response, CES salvaged vessels and became Crowley's in-house dive unit.

Eventually, six of these divers—John Graham, Greg Harem, Thom Davis, Mark Niccoli, Norm McCullum, and Joe Antonucci (a seventh, Tim Beaver, would not be far behind)—would break off from Crowley and form their own company, Global Diving & Salvage. The head of CES, Barry Paulson (affectionately known to his divers as "BP"), made Thom Davis his second hire in 1972. Davis was raised in rural southeast Missouri and had been diving since the age of 16. He taught himself how to scuba dive and began work as a teenager diving on odd jobs, such as finding sunk tackle boxes in ponds. Moving to Seattle, he worked for Professional Divers, Inc., in the Ballard neighborhood, one of the city's major maritime hubs, and did occasional diving jobs for Crowley Maritime. Then he went to work for Crowley Environmental full-time, and was soon joined by other young divers in their 20s. These divers had attended dive school, many at the Divers Institute of Technology (DIT) in Ballard. Diving was an opportunity to earn as much as a doctor or a lawyer, but without going through the motions at a traditional university, and offered what they thought to be a much more exciting career path.

Tim Beaver graduated from the University of Utah in 1974 with a

Top to bottom: The Alki Tavern: an important place in the history of Global Diving (photo by Katie Christensen); the Crowley Environmental crew (left to right): Thom Davis, Norm McCullum, Greg Harem, and Tim Beaver.

degree in English and an uncertain future ahead. What he did know was that there was a need in him that teaching English would not fulfill ("such as making a living wage," he said), and he wanted to be near the water. This latter desire took him to Seattle, where he began diving, and soon found his way to CES. "I thought, to dive for a living would be the coolest thing in the world," he said. After he graduated from DIT, much of Beaver's training was in the water, nagging his way into the dive rotation at CES.

John Graham started his dive career on a whim and received much of his post-DIT dive schooling in one of the world's foremost dive training grounds. "I met this buddy and he had been to diving school and was working in the Gulf of Mexico," he said. "I thought, well, I like the water, so I'll give it a try."

The Seattle-born Graham spent several years in and out of traditional universities before graduating from dive school and heading to the oilfields of the North Sea, located between Britain, France, and Scandinavia. "At that time, it was virtually impossible to work on the West Coast without experience," he said. "Going to work for the offshore oil sector in the Gulf of Mexico or North Sea was like going and getting your master's." Returning to the Northwest from overseas, Graham had the work experience to step in and become one of Crowley Environmental's lead divers.

Greg Harem also worked for CES but did not come to the company as a diver. He became friends with Thom Davis, and they made a plan to go scuba diving in Hawaii. Harem became scuba certified, and then gradually gained commercial experience through small jobs and courses at DIT, while continuing to work at Crowley. CES afforded the divers the opportunity to grow within their profession and in their individual careers, both in the water and out; the unit took their own calls, priced their own jobs, built their own relationships with clients, and compiled their own bills and time sheets. The diving outfit at Crowley Environmental had blossomed into a successful and essentially autonomous division. "They had it set up to where we did it all," Graham said. When work came in, the divers did everything short of typing up an invoice and collecting the check.

A shipbuilding boom at Todd Shipyard kept CES busy, including work for this icebreaker in drydock. Back, left to right: Mark Niccoli, Thom Davis, Joe Antonucci, Greg Harem, and Pete Shackleton. Front: John Graham, Paul Dougherty, and Norm McCullum.

Historical Values

GLOBAL DIVING & SALVAGE HAS BEEN A SUPPORTER OF MANY CAUSES over its history (see page 150), but one near and dear to the company's heart *is* history.

Global has been a supporter of the Puget Sound Maritime Historical Society and the Historical Diving Society, USA, organizations dedicated to preserving and promoting maritime history.

"We at Global carry on the tradition of diving, one that's got a very interesting, strong story as a specialized part of the maritime industry," Tim Beaver said. "It's valuable to remember we're part of this tradition, and it helps everyone be proud of what they're doing."

Leslie Leaney is cofounder of the Historical Diving Society, an organization whose efforts at "preservation through education" include an annual conference and a quarterly journal. Leaney, a London-born former music industry executive with bands including Guns N' Roses, has been able to count on Global Diving's support since meeting Tim Beaver at a trade conference in New Orleans ("He looked very hip with his cotton jacket, linen shirt, and goatee—I thought he was in the jazz band at first," Leaney recalled). Global has also sought Leaney's assistance in building the company's own collection of historical diving pieces.

"Global Diving's support has been vital to the stability of the Society," he said. "From my position as cofounder, they are the dream example of what an American diving company can do to encourage the preservation and promotion of the industry's history."

In addition to building their own collection of historical dive pieces, Global commissioned several paintings of company equipment and happenings (painting by Patrick Haskett).

"Crowley has to be credited with teaching us much of what we learned," Thom Davis said. "It was a great company to work for." According to Davis, Barry Paulson, who hired many of the divers, deserves credit for picking future industry leaders to round out his team. "The majority of the people Barry hired in the early '70s would become very successful. Several out of that group went on to become presidents and vice presidents of different companies. He had a knack for picking good people." Those people, aside from Global's founders, included Bruce Barto, who became a vice president at Crowley in 2001; Larry Pintler, a vice president for National Response Corporation (NRC); and Trygve Enger, who was a vice president for SEACOR Environmental Services before becoming an incident commander for NRC.

While talented, the 20-something-year-old men diving for CES had a tendency to occasionally act like, well, 20-something-year-old men. "Poor Barry," Tim Beaver recalled. "We used to drive him nuts." One such occasion—which Paulson somehow never caught wind of—occurred after a training exercise in Anacortes, near the oil refineries. The CES divers had been left to their own devices on board the 50-foot landing craft *Sag River* (which became the *MOP King* and was sister ship to Global Diving's *Prudhoe Bay*) while their supervisors and others in the training went ashore to debrief. In their equipment stash was a World War II-era line gun, an obsolete and never used tool that looked like a large-caliber pistol and was used to deploy lines between vessels. It had a large shield (like a TV satellite dish) to protect the shooter from blowback. Naturally, the CES crew had always wanted to see it in action, and finally, the right opportunity had presented itself. They tied a bumper, rather than a line, to the projectile so they could presumably retrieve it in the water.

Thom Davis was the shooter. "There was a big discussion about who was crazy enough to shoot the thing," he said. "It was my first—and last—time shooting it."

Davis aimed toward the shore and fired. The bumper went much farther than they anticipated—across the beach, across the road, into the woods, never to be found again.

The young Crowley Environmental team's salvage and oil spill work eventually turned into work at Todd Pacific and Lockheed shipyards, docking ships, inspecting docks, and patching World War II-era wooden drydocks that just couldn't seem to stay afloat. It was the relationships they built with these shipyards that would keep the young divers above water—just barely—after they decided to break off from Crowley and make a go of it on their own.

Frigates in drydock at Lockheed Shipyard, 1955.

chapter 2

On the Bottom

John Graham was still shivering at Terminal 128, even after he changed out of his dive gear into warm, dry clothes. Being cold is part and parcel of diving, but on this dive in the Duwamish River, the major industrial waterway leading south from Elliott Bay in Seattle, his equipment was no match for January temperatures that left ice floating in the water. His suit resembled a patched quilt, hardly the look for the owner of a dive company.

Yes, *owner*. Graham and five other founding members had in fact formed their own company, splitting from their former employer, Crowley Environmental Services (who had provided their gear before now). They pooled what resources they had: a dive hose, a helmet, several high-pressure air bottles (no compressor), and the Frankenstein outfit failing to provide warmth for Graham here on their very first job.

The owner-divers of Global Diving & Salvage were hired to patch the stove-pipe-thin hull plates of the *Raymond J. Bushey*, a small freighter moored far up the Duwamish at an out-of-the-way pier. Like any good divers, Graham and his fellow founders Mark Niccoli, Thom Davis, and Greg Harem persevered with the materials available to them. They came up with a method to patch on both the inside and outside of the hull. Battling the elements, they got the job done, and also got the company's first payment, in cash. It was the first income for a company that all hoped would generate enough revenue—just enough—to allow them to keep doing the thing they loved doing most: diving.

Thom Davis, above water, waits to assist diver John Graham.

"Nobody knows the bubbles we've seen."

JOHN GRAHAM

Discontent Amongst the Divers

For the founding six of Graham, Harem, Davis, Niccoli (known to all as Nick), Joe Antonucci, and Norm McCullum, the decision to start their own diving company was something all had been considering for nearly a year. Within Crowley, they were already essentially running their own business and calling their own shots.

That is, until the company hired an ex-Navy diver to manage the admittedly unruly group of brash young divers. According to Graham, the group's situation "got more autocratic and less democratic. That's when the thing started dissolving."

The divers, all in their mid-20s and brimming with the accompanying confidence (or arrogance) of their age group, bristled under the new guy's militaristic and tight-fisted rule. Talk began in earnest of trying something different. The pieces were there: Graham and Niccoli were hotshot divers; Davis could lead the oil spill and salvage efforts, and Harem was a quality diver and thinker; Antonucci knew how to pay taxes and do payroll; McCullum was a mechanic. They pestered another Crowley diver, Tim Beaver, to consider adding his talents.

Niccoli finally quit CES, and Graham was given his pink slip. Contacts at Lockheed and Todd Pacific shipyards, major clients of Crowley's diving team, came to Graham and Niccoli and encouraged Graham, Niccoli, Davis, and the others to start their own enterprise with the assurance that their talents would be needed.

Those six divers, young, single, assured of their abilities, and with nothing to lose, finally pulled the trigger. "I thought, If it doesn't work, I can always get another job," Graham said. "Jobs are easy to find." None of them, however, were under the illusion that starting a company was going to be easy. Few would have guessed that two years later they would be expanding, and Crowley would be out of the diving business.

Opposite: A diver at work.

Top: Global crews shovel corn to make ends meet.

Bottom: Earning the company's first paycheck on the Raymond J. Bushey *aboard the M/V* Bottom.

Taking a break on the ship's rudder.

Setting Up Shop

Once the decision was made to create a company, the process moved quickly. The fledgling business's first president was elected in an airport bar, minutes after touching down from a dive job in Texas for Crowley. Greg Harem was met at Seattle-Tacoma International Airport by the other five and sworn in, over drinks, before he even got to the baggage claim.

"I was the compromise candidate," Harem said. "There were a lot of disparate personalities in that group, and I think people felt that I could orchestrate some consensus and remain relatively fair and reasonable. I could mediate."

With formal leadership resolved, it was then decided that the company would be called Global Diving & Salvage because some of them had previously worked for a company called Global Transport Organization, and they liked the sound of "Global." Little did they know that the business really would go global decades later. They also did not want to limit their company from the outset. "We didn't want a name with 'Evergreen' or 'Northwest,'" John Graham said. "We didn't want any limitations, but who knew what the future would hold."

The owners each bought in to the company, pooling their limited savings for insurance, their biggest initial expense. Several thousand dollars of insurance money was kept in a coffee can in Thom Davis's home freezer, "where we thought it would be safe," he said.

One member of the Crowley diving team who was offered a stake but did not immediately join was Tim Beaver. "I couldn't afford to take a flier like that," he said. "I was the only one that was married and had a house payment."

In the coming months, there would be opportunities for Beaver and others to contribute to Global's upstart efforts, but for now the founding divers just needed to get to work. And in a down economy, work was not easy to come by.

Shipyard Work Starts and Saves Global

For much of Global's infancy, the divers spent their days impatiently

The M/V Bottom *in all its glory.*

Left to right: Don Hibbs, Henry Nakagawa, John Graham, and Thom Davis inspecting the Chittenden Locks, Lake Washington Ship Canal, Seattle.

Greg Harem in the decompression chamber.

waiting by the phone (a landline in those days), praying for work. The company had rented an office—more precisely, the front room of what was originally a World War II barracks with an office in front and room for bunks in the back—on Harbor Island from Bob Rausch, a marine junk man who lived upstairs and whom they had met in that very space years earlier. The wood-frame building at 2763 13th Avenue Southwest, set on concrete blocks and originally meant to be temporary, was "the biggest fire trap you can imagine," Tim Beaver said. In what would be the first of several circuitous twists of fate in the company's history, the very same barracks/office had previously housed Crowley Environmental Services, then known as MOPS, before they outgrew the space.

Rausch helped their entrepreneurial efforts, leasing Global the company's first truck (an ex-Navy four-wheel-drive flatbed still in the original military gray) and much of their rigging tools and equipment, and later even going so far as operating as their de facto banker when they needed a bond for a job. Still, he couldn't make the phone ring.

One man who could and would call time and again was Dave Anderson, dockmaster at Todd Pacific Shipyard. Anderson had met the Global team while they were diving for Crowley, patching wooden drydocks at Todd in an admittedly futile effort to keep them afloat. After hours of scraping barnacles and other growth overhead with a flattened hoe, by hand, they rolled out yellow PVC canvas (a Visqueen-like plastic sheeting) and tacked it down with staple guns. Then the divers, working upside-down underwater, used four-pound sledgehammers to drive, by hand, thousands of 16d stainless steel ring shank nails through 4-inch-wide by 8-foot-long creosote-treated strips of wood (battens) and into the drydock. "It was like working on the ceiling of your living room," Tim Beaver said. "We did that week after week, month after month. It was hard work."

Their perseverance was impressive. It became apparent to Anderson that he did not want to work with any other divers. "These guys learned so quickly," he said. "They knew what I wanted before I even knew what I wanted. They'd tell me of a problem before it arose: 'this is what we're finding, we suggest this, what do you want to do?' "

The Allure of Diving

A diver keeps his nose to the grindstone.

"DIVING IS NOT GLAMOROUS," said Global Diving president Devon Grennan. "It's not like you're Clive Cussler's lead hero setting explosives underwater, swimming away as it blows up to a sandy beach with a beautiful blonde in a bikini. It is you, 100 feet down next to an oil platform, trying to weld X amount of linear feet in 45 minutes by yourself, and you're really cold. And if things don't go exactly right, you might not go home."

Diving is lonely, frequently frustrating, and inherently risky—so why would anyone want to do it for a living, especially when the job takes divers away from their families for weeks at a time, and can demand their presence at any hour on any day?

The compensation is certainly an important reason. The telltale sign of a successful, significant long-term project—for company and diver alike—is to see a small fleet of brand-new pickup trucks in the office parking lot after the job's completion. Still, the attraction has to extend beyond new trucks; diving is a passion, and many of the best divers cannot imagine themselves working in any other profession.

Jackie Lewis, after her decades working in close quarters with divers, thinks she knows why. "They're attracted to the adventure, the fact that not a lot of people can do it, and the money," she said. "And it is dangerous."

The danger is an attraction, particularly for young males beginning their professional careers. Grennan likens the diver personality type to that of a race car driver, who finds the risks of his or her profession thrilling. The risks also create exclusivity—Grennan estimates there are 3,000 professional divers in the United States, and "maybe five percent excel at their craft," with Global employing a good portion of that top talent.

Greg Harem, a former diver, agrees with Lewis's take on diving's allure, but also enjoyed the camaraderie. What truly hooked him on diving, though, was having the weight of a project squarely on his shoulders. "The greatest thing for me was that when you're underwater, you're the person," he said. "Everybody on the surface is working to make you more effective. You have to make observations, use your ingenuity, and work with the topside crew to come up with solutions. But while you're in the water, the process goes through you, and it's a great feeling when you know the client trusts you to do your job and to provide him with the information he needs to do his job."

Dan Gilchrist, one of Global's lead divers, has been a professional diver for 25 years, a remarkably long tenure in the water. He credits his physical longevity to a commitment to proper diet and exercise—"Things start to change as you get older and you've got to change with it," he said—but what allows him to mentally handle the day-in and day-out rigors of his profession is harder to pin down.

"I really don't know how to explain it," he said. "I guess it's the passion for dive work, professional pride really, and being driven to go do something and make it the best it can be by giving it all your best. It's my career, it's my job."

Dan Pierson prepares his hot water suit for a cold-water dive.

Anderson had encouraged the creation of Global by assuring work, and with that, the divers came calling. "They didn't have anything," he recalled. "They were in tough shape. But I told them, if you guys can hang on, there's a lot of work coming up." The work he referred to was Navy shipbuilding projects awarded to Todd and Lockheed, the result of efforts by Washington's longtime Senator Warren G. Magnuson to bring shipbuilding work back to the Northwest. When the project began, the shipyards hummed at levels not seen since World War II.

Until then, the divers jumped at any chance to work, even if it meant taking plastic sheeting and a hand maul to that same drydock for the umpteenth time. Anderson could depend on the divers and expect their prompt help at any hour, day or night. "You didn't know what they were doing or what it was interrupting, but you could call. That fast response has always been a trademark at Global." Anderson, who retired in January 2012, estimates that in three decades of working with Global, he never had to wait more than an hour for a diver.

The work at Todd was crucial for the company's survival. They simply were not known yet in the maritime industry. The divers scraped by. They rotated the daily lunch duty of wearing an old, musty, oversized jacket to Shakey's buffet on West Seattle's 35th Avenue, stuffing its many pockets with chicken and rolls to be shared back at the office.

Then Todd's huge contract building frigates for the Navy started, and work became steadier. Global also began similar work at Lockheed Shipyard upon a recommendation that Anderson made to Dockmaster Pat Majors. With these work opportunities came chances at bigger, more complex jobs that tested the owner-divers' capabilities—and their tolerance for risk. On one such job, a complicated dry-docking that involved two six-man teams, with two men each in the water, working simultaneously on the same ship (one team at the stern, one at the bow), many at Global were unsure they could round up enough qualified divers for the job, let alone do it.

In the Harbor Island office, a lengthy talk led to naysaying and doubt as thick as the cigarette smoke enveloping the young divers. "We were talking about all the things that could go wrong, and how it was going to ruin our reputation," Tim Beaver said. "We were just about ready to call the shipyard and cancel the thing."

No one noticed, but John Graham quietly got up and left the doom and gloom. Soon the others heard crashing and banging outside. They walked out to the loading dock behind the office to find Graham throwing dive hoses into the back of his truck. "We're doing this goddamn job, and I'm tired of talking about it," he growled. He continued to load gear, and the others followed suit. Global then drove to the shipyard, did the job, and did it successfully—no reputations ruined, only enhanced. It was a turning point for the fledgling company: as uncomfortable as it may have been, they would continue pushing their limits to keep getting better, steadier, more exciting, and more lucrative projects.

Todd Shipyard humming with work in the early 1980s.

As they got more work, "you could actually watch the divers equip themselves," Dave Anderson said. Gone were the ragtag patch jobs, replaced by sturdy new neoprene drysuits, new Aquadyne band masks with lights, and new dive radios, all stored in fancy mahogany boxes with a brass handle on top. "All of a sudden, they were coming out of the 19th century."

The divers themselves weren't seeing the benefits in their personal bank accounts. They each took home $75 a week; the rest went back into equipment. They may still have been cash-poor, but at least they were now patch-free.

chapter 3

Travel Lines

The members of the original Global Diving team are quick to concede they did not fully comprehend what they were getting into by starting their own company. Several declared that knowing what they know now, they wouldn't take the same risk. It's safe to say they were in over their heads through much of the early years, blissfully ignorant of the magnitude of the professional and financial risks taken on individual projects and in business in general. On one particular job in Anacortes, Washington, in 1984, pouring concrete pilings for a Texaco oil refinery, the challenge was all too apparent.

Global was subcontracting for Manson Construction, one of the Northwest's most respected and longest-tenured names in marine construction. They had been bugging Tike Hillman at Manson to give them a shot at diving as a subcontractor on one of his big projects, and this was it—but as they later found out from the gruff Manson supervisor John Molvar, only because the previous dive subcontractor could not complete the job. Global submitted a bid to Manson, which Hillman suggested they revise, upward, twice, so they would not forget anything. "He wanted us to succeed," John Graham said, "and he wanted us to make him look good."

Hillman had one caveat: Global team members needed to be in the pile drivers' union to qualify for the job. The Brotherhood of

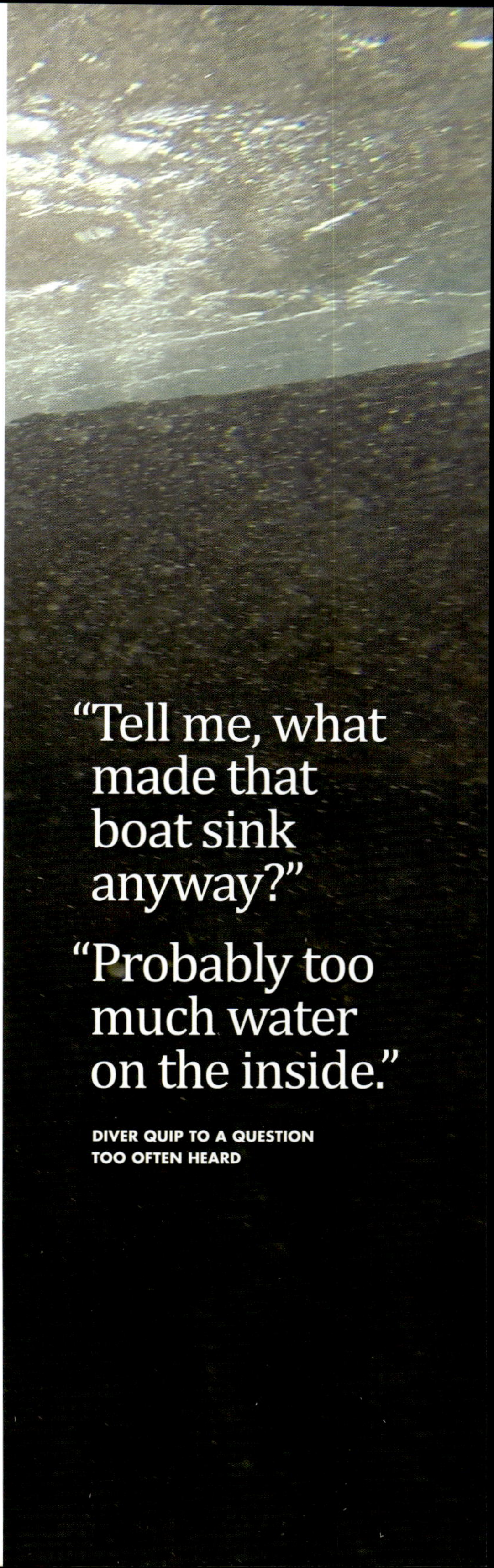

A diver cuts apart a ship in shallow water on the Palmyra Atoll project.

Gary Griffith (left) and Thom Davis aboard the Bottom *on the Texaco Refinery project in Anacortes, Washington.*

Carpenters, however, thought they had more than enough divers in the hall for the work available and were not interested in adding any more. After being stonewalled by the union representative, the Global leads returned with a highly recommended local labor lawyer in tow. The chain-smoking business rep Bill Sullivan was so flustered that they had brought this attorney with them that he accidentally put fire to the filter end of his next cigarette, but Global walked out with an agreement.

After clobbering coffee-cup-sized barnacles and rock scallops off the steel H-piles with a 5,000-PSI water blaster, it came time to pour concrete and shape the encasements. Timing was of the essence in getting trucks loaded with wet concrete down a two-mile, one-way dock before the concrete set.

The divers, all owners in the company, often diving twice a day, wrestled underwater with finicky nylon bags. The bags, called Sea Form Systems, were a new technology to them. Though they had practiced with the new systems earlier, "we were still learning how to do this stuff," said Tim Beaver, who brought the job to Global and had by now bought into the company (more on that later). The actual installation process turned out to be a lot tougher than anticipated, particularly when it involved yards of wet concrete in a sagging form on a batter pile (an angled piling added to help horizontally stabilize a structure). "We were definitely on a steep learning curve," Greg Harem said. The diver would zip the bags around a reinforcing bar cage placed around the piling, then tremie concrete with a pump, continuously hammering the bloated form with a 20-pound sledgehammer to shock it into shape. Global bid the job with Sea Form, rather than the traditional hard form, in an effort

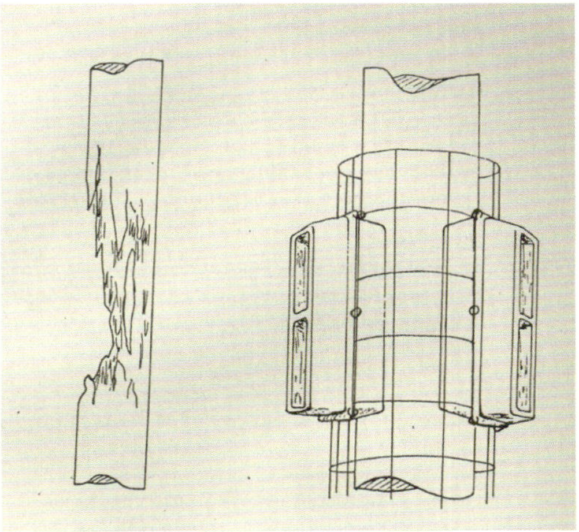

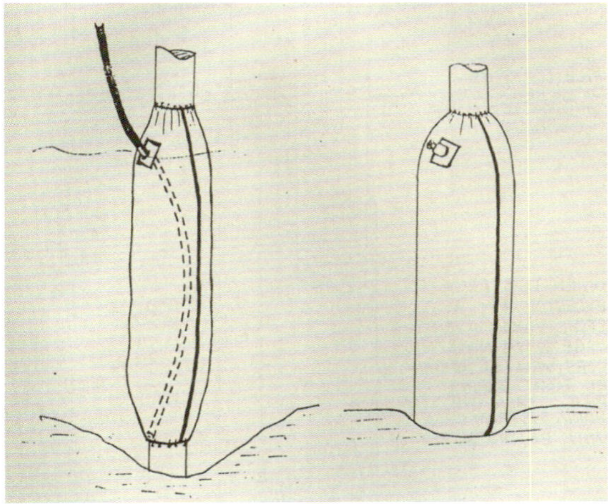

On the Texaco Refinery job, before a finished Sea Form System could support any structure (top), the diver had to battle with the "time saving" nylon bag. First, a reinforced bar cage was placed around the rotted or damaged pile (left). Then, the bag was zipped up, concrete was poured in, and the Sea Form had to be quickly shaped with a sledgehammer (right).

to save time and money; this was an early example of Global's use of outside-the-box approaches to tackle problems and deliver better value to the client.

It wasn't saving time. The divers were fighting a losing battle against slumping concrete. "You might get up six or seven feet and realize you'd gone about it the wrong way," Harem said. "The only choice was to unzip and dump everything. Then you had to go down again and blast the concrete away from the pilings so you could clear it and start over, all before it hardened up."

Frustrated and exasperated, the Global team pushed on. They refined their craft as they went, and working into the icy winter months, finally completed the encasement work to Hillman's satisfaction. "We learned a lot on that job," Graham said. "How to deal with a bunch of different clients, the union aspect, working under a prime contractor and keeping them happy. Dealing with the refinery, and all their rules and regulations. All in addition to honing our craft."

The Texaco Refinery job led to three more similar encasement projects. They were the company's most lucrative projects to date, worth more money than the company made in its entire first year. In 1980, the company's operating statement went on to a second page only to impress banks. Now in its fourth year, after Texaco and the three subsequent encasement gigs, Global was able to buy its first truck off the lot, a brand-new gray 1984 Ford 4x4 diesel F-250.

"No Going Backwards"

After a trying first year, Global Diving & Salvage had established itself as a reliable crew of skilled, determined professionals. They had a particular way of doing business, influenced by their own experiences and observations of other service providers on how to run an operation, and often more important, how not to.

Top to bottom: Greg Harem in Anacortes on the Texaco Refinery job; Global's ex-Navy flatbed truck loaded and ready to go; Tim Beaver (left) and John Graham raise a crashed Lake Buccaneer seaplane in Long Lake, Washington.

Global founders bolt the company sign to the Harbor Island office in what Thom Davis called his "proudest moment at Global."

Salvage

THE END GOAL OF A SALVAGE EFFORT is to get something back out of the water that shouldn't be there in the first place, usually a vessel and/or its cargo. How it's done depends entirely on the situation. To raise a vessel, salvors may use pumps placed underwater, or may lift with a crane by attaching cables. Retrieving cargo today may be done using remotely operated vehicles (ROVs) with grasping capabilities.

Regardless of the method, what is most important for a diver on a salvage job is to be alert, even more so than usual. Salvage work requires a different mentality than a construction project. Salvage divers may be entering unstable and deteriorating environments, and so must recognize when to press forward with a task, and—more important—when not to. In recent years, advances in 3-D rendering technology, ROV-provided video, and still images have helped improve safety on salvages by giving divers a better game plan before going under.

Crewmembers look on as a sunken boat is raised by crane in Lake Union, Seattle.

One early rule: before any of the owners saw a dime, all vendors, bills, insurance, and subcontractors were paid on time. Too many waterfront suppliers were wary of dive companies because other outfits, in particular a company called Garrison 8, went bankrupt, leaving many vendors and suppliers empty-handed. As such, trust and credit had to be earned by action and over time.

Delivering a quality product certainly helped. Clients of diving companies rarely see the work they pay for. Integrity was essential, as was communication. Global delivered thorough reports to clients, working late, typing observations and findings, and painstakingly cutting and pasting photos taken underwater.

For the divers themselves, respect and safety were paramount. "Treat everyone the way you want to be treated" was a golden and enduring mantra. After losing two colleagues to accidents while at Crowley, the founders put a premium on job safety and preparedness.

The company's reputation grew, with the evidence visible at the Harbor Island office. They slowly took over more of Bob Rausch's building, room by room, and made it their own. Thom Davis's proudest moment at Global came when they bolted a sign to the front of the building after the company's first year. It featured the company's logo, designed by Bob Allen, a friend and semi-professional artist. The gold dive helmet, sitting just above oil-covered water at the emblem's center, featured a mistake: there was a missing wingnut and stud. The divers liked it just the way it was.

"It represented a working helmet," John Graham said. "Sometimes on jobs you have to make do with less-than-perfect equipment. To me, it said, We're going to do this until we get it done. If we have to make a hammer out of a rock and a stick, then that's what we'll do."

Above: The original office, complete with chain-link railing and the company anchor.

Left: Tim Beaver in his office. To his right is the infamous green barber chair, which is still a fixture at Global's Seattle office today. The chair was originally borrowed from Bob Rausch's warehouse of eclectic surplus gear, which included such items as two pallets of straitjackets and a 55-gallon drum of military watch oil.

That attitude was evident in the handrails newly erected on the front steps, made from anchor chains welded together. They painted the building black and white, subbing the darker shade for the original blue. A brass nameplate was affixed to the front door. The signage announced to their industrial and maritime neighbors on Harbor Island—including Todd Pacific Shipyard, Rainier Petroleum, Shell, Texaco, and their old employer, Crowley—that they were open for business, and planning to stick around for a while.

After a year and a half, the founders felt comfortable enough in the company's income to be able to make a commitment to Tim Beaver, their Crowley cohort. Beaver had worked for Global on the weekends and as opportunities arose. In the meantime, he quit Crowley and was working as a tankerman, diver, and boat operator for various companies including Timmerman Marine, Foss, Crowley, and Pacific Northern Oil. Soon, he and his father, Jack, started a small company, Beaver Pacific, Ltd., and would hire on Global as a subcontractor. Using assets from the venture, Beaver was able to buy in to Global Diving, but only after the current six determined that they had the resources necessary for him to take a salary.

"We decided that if a guy was coming in as a partner and he had a family, we would not let him go backwards," John Graham said. "Those people got regular paychecks. The ones of us that were single, we got some dough when there was extra dough."

Extra cash was hard to come by. Able bodies willing to work for the love of diving were not. "We had a good little support group of semi-employed people," Tim Beaver said. Many became acquainted with the Global guys at what the divers referred to as their "second office," the Alki Tavern. Divers were known around the offices by nicknames: Agent Orange, Doctor Demento, The Cobra. Each nickname had a story behind it, though most of them remain elusive to historical record. One semi-employed type was Steve Eyler, better known as Four Buck, for the wages he was originally willing to accept per hour.

Nicknames were also affectionately bestowed on gear. John Graham's yellow and black Boeing surplus Datsun pickup became known as The Bee, for its paint job, but also for its muffler-less drone; Greg Harem's beat-up pickup was called The Crummy; and the company boat became

Global built its Harbor Island shop and warehouse in 1985.

"Stormin' Norman the Oil Spill Foreman," Norm McCullum.

Some of the semi-employed types included (left to right) Tim Galyon, Jeff Eyler, Mike Langen, Pete "the Shack" Shackleton, and Steve "Four Buck" Eyler.

Roger, in homage to a three-legged dog met on a job.

A diver who came through the door in the early years was Mike Langen, who dove seasonally in the Gulf Coast for offshore oil companies. He was searching for work that would keep him in Seattle full-time and hoped to find it at Global.

"There was a vibe about that place that I really liked," Langen said. "You walked in the front door and there was the equipment room; no receptionist, nothing else. There was something about that that made me think, I could work at this place." Langen's persistence would eventually pay off.

Manson's Hawken *begins to lift the* Warrior.

Wrestling with the *Warrior*

In the meantime, what work there was went to the owner-divers. They worked in two-man teams, all diving and tending, making frequent trips up and down Interstate-5 to wherever jobs could be had. They pulled all-nighters in Tacoma doing work for Sea-Land, replacing rivets, installing sea chest blanks, welding up cracked rudders, and even scrubbing hulls on ships as they offloaded.

The Global team truly did take work where they could find it—one early job involved shoveling corn from an overturned semi—but in addition to marine construction, such as the Texaco Refinery job, they also eventually branched out into casualty response.

While some of the crew continued work in Anacortes mastering the art of Sea Form Systems, others left to work on the salvage of the *Warrior*, a derelict vessel that had sunk near Blaine, Washington. The *Warrior* lived up to its name, not leaving its resting place without a fight. Contacted by the sunken vessel's owners first, Global's young divers brought the crane job to Manson. However, the grizzled Manson managers at the time decided that working under junior statesman Global Diving would be like having the tail wag the dog, so therefore they would be calling the shots. After they ignored strong suggestions from Global divers to take the time to place timber-softening pads under the wires to spread the load on the aged hull, bare wires were set underneath the vessel. Manson's biggest crane,

Greg Harem stands in front of the Warrior*, at the vessel's new resting place.*

Heavy Lifting

THE *WARRIOR* SALVAGE WAS DIFFICULT in its own right, and sometimes, one crane isn't enough.

For the heaviest lifts, Global has successfully brought together fierce competitors to rent their equipment and loan their expertise to the toughest jobs, ones where multiple cranes are required for a stable, safe raise. Global first used cranes from both Manson and General Construction on the 1994 salvage of the *Jupiter*, a 117-foot steel fishing boat that ran aground and rolled over in Seattle's Shilshole Bay near the Ballard Locks.

"There wasn't a crane anywhere in the Northwest big enough to do the job on its own. We needed two big cranes, and Manson and General Construction had the only two cranes available at the time," Tim Beaver recalled. "They were willing to work together under our leadership because they trusted us. They were staunch, staunch competitors, but we had worked with both companies for years and they knew us, our people, and the quality of our work."

Other noteworthy projects have required multiple cranes from multiple companies, including the salvage of a wooden tug that sank near the Ballard Bridge in Seattle. A derelict tug in Poulsbo, Washington, had to be lifted by two cranes, one from Manson and one from General Construction, before a drydock could be slid underneath it.

the *Hawken*, attempted to raise the *Warrior*, despite the Global crew's warnings that the wire would slice right through the rotten wood hull planking bolted between steel framework. "We're pumping water, they heave up on this thing, it starts coming up, and slowly one end of it quivers and drops," Beaver recalled.

In between its steel frames, the wood planking splintered into shark's teeth covered by a foot or more of mussels and barnacles. However, the *Warrior* had an unusual concrete deck, which in its better days had enabled forklifts to move immense newsprint rolls; for years the vessel carried black oil to the outlying paper mills in Puget Sound and newsprint back to Seattle. The salvage team agreed to keep lifting so the wire would eventually stop at the concrete deck, and then they could raise it. It worked, but then the question became, Where to set it down for the time being?

Two cranes were needed to lift the M/V Florence Filberg *out of the Lake Washington Ship Canal, Seattle.*

Nearby land was vacant, just old cannery buildings. But because it was destined to be the future site of the prestigious Semiahmoo golf resort, the owner of the land was not going to permit this blight on his significant investment. He called the police. The police officer came to the dock, looked up to see the vessel high above the water, lifted by a 10-story-high crane, and pointedly offered his judgment: "If they want to pick this thing up and put it into the middle of Main Street in downtown Blaine, there ain't much I can do about it."

In spite of it all, the *Warrior* salvage furthered Global's relationship with Manson. "They started listening to us more," Beaver said. "Now when we go and do a big job, we work very well with them."

Global's work on the Texaco pilings and the *Warrior* raised its profile on marine construction and casualty response jobs. The owner-divers and all the divers working out of the Harbor Island office worthy of a nickname were steadily gaining credibility. However, it would be the company's work on oil spills that would significantly raise its profile, and bottom line, by bringing Global into the lucrative business of regulatory work.

PHILADELPHIA

chapter 4

Cleaning Up

New Year's Day 1986 found a Global Diving crew adorned in Hawaiian shirts and barhopping in Port Angeles, Washington. John "JG" Graham, "Tugboat Tommy" Thom Davis, and "Stormin' Norman the Oil Spill Foreman" Norm McCullum led a gang of Global's beloved semi-employed types including Ronzo, Semper Fi, the Pointer, and Lance Romance around the small, remote logging town on the north shore of the Olympic Peninsula. Their tongue-in-cheek tropical regalia clashed with the typically gray, cold, and wet Northwest winter, and their carefree celebration was a stark contrast to the somber carnage they had been dealing with for the past couple of weeks.

On December 21st, the tanker *ARCO Anchorage* had turned too close to nearby Ediz Hook, a sand spit extending from the Olympic Peninsula's northern shore, and ended up in shallow water. A rock ripped through the inch-and-a-half-thick steel protecting portside tanks, spilling 239,000 gallons of Alaska crude oil, setting a not-so-prestigious record for spilled oil in Washington State. With a National Wildlife Refuge just 14 miles away at Dungeness Spit, early responders decided to protect the Spit first and foremost, which left Ediz Hook to take the brunt of the spill.

Soon after, Global Diving got the call to assist in the difficult cleanup. The contained oil along the sand spit sank into the sand. Crews worked long hours, but with every tide more oil came back up.

"Our Business Is Booming."

GLOBAL COMPANY ADAGE FOR OIL SPILL WORK

Thom Davis aboard the MCN 5 *in the Guemes Channel of the San Juan Islands, Washington, during a barge salvage and oil spill project.*

ARCO ANCHORAGE

The company's oil spill response service, headed by Thom Davis, had become known for its innovative tactics to combat these unexpected environmental disasters, and this spill only added to its reputation. Davis, working with an ARCO representative, devised an ingenious method to get the oil out of the sand that would become known as the "muck monster technique."

"He took a Caterpillar tractor with a big side boom on it, put a hydraulic pump out on the boom over the water, and then ran pipes down the back of the Cat to each of the ripper teeth," Tim Beaver said. "As the tide came in, they'd run this tractor down the beach, practically underwater itself, and it would pump salt water into the sand and out would float the embedded black crude oil.

"Everybody thought that was crazy. But when we did that for two or three tide cycles, the beach stayed clean. Basically the only other option was to haul the whole beach away. Anyway, it worked."

The Global team, dressed for sun-and-sand in a region of evergreens and constant cloud cover, had reason to celebrate that New Year's. Not only had they found a solution to a key issue in the response, but the company had also survived the first round of cuts, so to speak. On oil spill responses, many hands are hired for the initial cleanup efforts, but only a select few stick around to see the project through completion. Global would have work for months ahead.

"The bread and butter were the everyday diving jobs, but the gravy was the oil spills," said Jackie Lewis, who would join the company later that year as bookkeeper and become receptionist, human resources, administrative assistant, compliance officer, and all-around "mom" of Global Diving & Salvage for decades to come.

Key Additions to the Global Family

In the mid-1980s, Global embarked on a brief venture in tugboating with the purchase of the *Spike* and, later, the *Spar* (see sidebar, page 53). Overall, it was

The ARCO Anchorage *spill called for an intensive cleanup effort. Top to bottom: Global cleaned and retagged 14 million board feet of premium lumber that had been drenched by oil in the spill; the cleanup crew breaks for lunch.*

Oil Spill Response

NO OIL SPILL RESPONSE IS EVER THE SAME, thanks to factors like wind, weather, distance from shore, and the amount spilled. However, there are some tried-and-true methods crews use to contain and clean up these literally and figuratively messy situations.

When certain kinds of oil hit water, they float on the top and spread out in a thin layer. Responders can place preventative booms (floating barriers with a containment skirt hanging underwater) in the water to contain the spread, then either "skim" (suck up) the oil, soak it up with floating sorbents, or in a worst-case scenario, burn it off. Spill containment booms come in many different shapes and sizes and are frequently used in responses, giving credence to the Global Diving adage "Our Business Is Booming."

Once oil reaches the shore, the game changes, becoming more complicated. Through the decades, beach cleanup has been Global's forte. Oil may wash ashore as small, black tar balls, created when oil mixes with sand and rocks, or as a thick muck across the beach. Global has innovated techniques throughout the years—including the "pom-pom lines," or oil snares strung together, and the "muck monster" technique—to improve industry-standard methods of onshore cleanup. Further, Thom Davis and his crews expanded the use of all-terrain vehicles (ATVs) to work with helicopters and small landing craft. This helped make more efficient the backbreaking work of removing contaminated driftwood, seaweed, and sand from polluted beaches from Alaska to California.

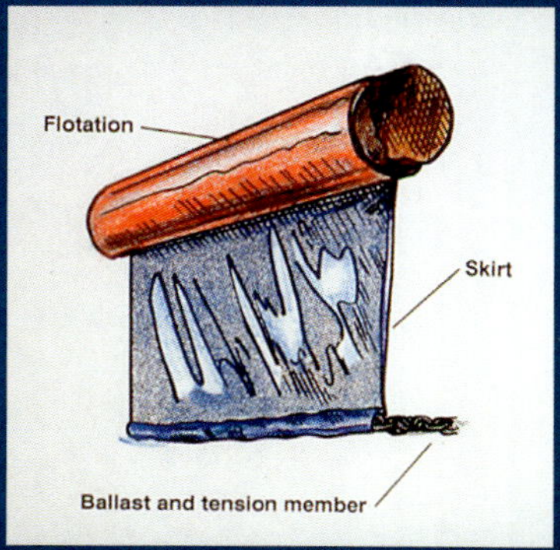

A crewmember sets boom in Anacortes, Washington.

Left: Underwater, a skirt is attached to the floating portion of the boom to contain any oil leakage.

not worth the trouble—"Looking back, I would've fired myself," Tim Beaver said—but it did lead to the hiring of mechanic Gary Duff. Global had previously contracted Duff and his boat, the ex-*Ivar Foss*, to assist on jobs. On one of their very first jobs together, a quarter-inch compression fitting broke and Duff, armed with extra pipe fittings, fixed it, allowing work to continue with minimal interruption.

The Global team was impressed, and asked Duff to interview for a position over Sunday breakfast. "They said, 'Tell us what you can do,'" Duff recalled. "I thought, they're not going to believe what I tell them."

Duff became a self-described "gear head" at the age of 13 when he found an interest in amateur radio and began fixing neighborhood Doodle Bug scooters. A graduate of Edison Technical School, he spent time as a chief mechanic at the Indianapolis 500 before buying his own tugboat.

"He is an uncannily smart guy," Tim Beaver said. "It all goes in and it never comes out. He remembers the exact kind of rubber thickness you need to build a specific gasket for a specific part, and every little vendor in town that sells this particular custom spring."

Even after he fixed the Crummy's blown-out turn signal levers within a week of joining Global, not all were convinced of Duff's prowess. Later, after a transmission blew out in the company's Chevrolet van, he took it apart in the shop. Thom Davis walked over to see the massive mosaic of nuts, bolts, screws, and gears of a Turbo-Hydramatic transmission haphazardly strewn about the shop. He returned to the office with his report: "That truck's never gonna run again." Three hours later, Duff rolled up in front of

Right: Gary Duff keeping it all running in the Global yard.

the office in the truck, running like new, and asked if anyone wanted to go for a test drive. His talents were never questioned again.

Besides fixing whatever fell into disrepair on larger machinery, Gary Duff improved equipment for the divers. He rebuilt the air compressors to make the air taste better, or "sweet" in diver lingo, by honing (or finely etching) the cylinder walls and seating rings so that oil would not be dragged up into the compression part. Even after Duff eventually moved on from the company, he would still come back for years to show new mechanics his methods after the divers had pestered them for Duff's sweet air. He also used his radio expertise to improve sound over the dive radios, meticulously transplanting parts from RadioShack and adjusting impedance to create communication that was much less crackly and crunchy. Duff became a fixture at Global's shop and could often be found hard at work in his standard daily work uniform of a Hickory shirt, Levi jeans, and Red Wing foam sole tugboat slippers.

Jackie "the Queen" Lewis.

Another Global mainstay first arrived at the Harbor Island office in 1986, when Jackie Lewis was taken on a quick tour of the humble facilities by President Greg Harem. They had already met earlier at an employment agency in Renton. Lewis was selling real estate part-time but had heard from a friend at the agency about an open bookkeeper position, for which she was to meet the company president. She wore a skirt and jacket, which impressed Harem; he thought her well-kept appearance could not help but set a good example in the casual (at times to a fault) office environment, amid the dust and din of Harbor Island. After talking for all of two minutes, Harem offered her the job, but admitted, "In fairness to you, you need to come to Harbor Island and see what you're getting into."

The tour was quick—there wasn't a whole lot to see, just an equipment room, combination bathroom/kitchen/coat room/storage area, and the dive locker—and very informative.

The *Spike* and the *Spar*

Global Diving tried its hand at tugboats with the purchase of the *Spike*, a mothballed YTL (Yard Tug Little, a small harbor tug) built in 1944 for use in World War II. The idea was to have in-house tugs available for the growing amount of salvage and diving work that Global was taking on during the 1980s. Tim Beaver had experience with tugs during his previous venture, Beaver Pacific, but it was the hiring of Gary Duff and his knowledge of all things with a motor that helped bring the *Spike* back to life. Global bought another tug, the *Spar*, to help with the second season of recovery efforts for the *Exxon Valdez* oil spill in 1990.

Global's "tangent into tugs," as Beaver called it, "was not the smartest thing we ever did, but we all learned a lot from Gary Duff about seamanship, vessel construction, and the perspective of the vessel owner."

"The tug thing was a bad deal, in a way," Duff said. "They aren't simple machines; you have to have your wits about you."

Most divers were uninterested in learning how to operate tugs. "Tugboating is a complicated, detail-oriented business and you have to love it," Beaver said. "Put a bunch of divers on the deck of a tugboat, and you're not going to get the results you want. We finally understood that and got rid of the boats."

By the mid-1990s, Global had sold both the *Spike* and the *Spar*. Global's greatest success with these boats came in 1987, when the Mighty Tug *Spike* took first place in the Olympia, Washington, Tugboat Races.

More recently, Global Diving has triumphed in building company morale by participating in the annual Tugboat Races during the Seattle Maritime Festival, touted as the biggest tugboat race in the world. The company has been a repeat winner of the Tugboat Race Chairman's Award (known company-wide as the Red Solo Cup Award), given to the boat having the best time, thanks to Global merrymakers on board the *Prudhoe Bay*, a utility vessel Global bought in 1992 and retrofitted in 2008.

"Dive underwear smells like nothing in the free world," Lewis said. Disgusted but undeterred, she took the job. "Global is a company you really get caught up in."

Lewis quickly made her mark. The only female employee for a company in a high-testosterone industry, she proved to be tough but loving, one of the boys in the dive locker and the face of the company to insurers, banks, and other partners. "She's an incredible woman," Gary Duff said. "A lot of class, but you don't mess with her. She is a huge part of why the company is the way it is." Lewis went to job sites to do payroll, paid bills, bought insurance, fetched sandwiches at Joe's, and helped pick out Christmas gifts for the wives and girlfriends.

Bruce Humberstone, right, with John Graham.

There were few limits to the tasks that Lewis took on, even repiercing the ear of Global's first full-time employee with a sharp earring, recalled Bruce Humberstone, who in 1988 was the first Global diver to be guaranteed 40 hours a week in shop time with benefits. Humberstone studied at DIT and had been working on and off for Global for several years, splitting his time between Seattle and Houston, where he still dove offshore. Through a mutual connection he had been put in touch with Global, who called upon him while he was in the Northwest to help with various projects before the company could make a full-time commitment, and Humberstone could stay year-round in his native Washington.

"At Global we'd do a couple jobs a week back then, which was different for me," he said. "In the oilfield, we did major projects, three-weeks or month-long. Here it was entirely the opposite: put two bottles in a Whaler and go look at a boat. It was more of a family-type operation, which had nice aspects to it."

The *Nestucca*

Frequently, the Global crew would clean up small oil spills right off Harbor Island at the shipyards, oil terminals, or container facilities. In those much less regulated times, the crew would respond in sandals, shorts, and T-shirts. After the job was done, Jackie Lewis would walk over with a roll of paper towels and root beer floats.

Soon, the spills would be much bigger, but the Global family remained just as close-knit. The *ARCO Anchorage* response was excellent practice for the *Nestucca* barge spill, almost three years later to the day in 1988. On midnight of December 22, just off Grays Harbor, Washington, the towline from the tug *Ocean Service* connecting to the *Nestucca* snapped. In its efforts to recapture the drifting barge in severe weather, the *Ocean Service* punctured a cargo tank. Over 227,000 gallons of fuel gushed into the bay, just north of the mouth of the Columbia River.

The spill called for a massive response. Global Diving enlisted the help of the nearby Quileute and Makah tribes, hiring some 30 workers to assist. Jackie Lewis did payroll out of her hotel room, where there would be a line from her door down the hallway on Friday paydays.

Lines of a different kind helped make the response a huge success, and represented an important advancement in marine cleanup efforts. By accident Thom Davis devised a now-common cleanup method in which oil snares are fixed together to a rope for easier retrieval. Oil snares, known as "pom-poms," are commonly used today in spill response to absorb oil. In an effort to keep the pom-poms from migrating to another beach during tide change, Davis had strung

Top: Tim Beaver and Greg Harem (middle, not facing camera) eat sandwiches with other crewmembers on the beach.

Bottom: A demonstration of a pom-pom on a snare line.

them together off the ground. The line broke. "The snare washed around the beach. We were surprised when we went back and saw where they had scrubbed the rocks clean of oil," Davis said. "We just stumbled onto it."

With oil snare lines, the cleanup crew could leave strings of snares across the beach. When the tide came in, they sopped up oil, and they were easy to collect afterward. Global rented a warehouse, hired workers by word-of-mouth from a local restaurant, and made oil snare lines as fast as they could.

Exxon Valdez

The "pom-pom lines" became even more valuable when Coast Guard officials came looking for help on a disaster of much larger scale. The 987-foot tanker *Exxon Valdez* ran aground on Bligh Reef in southern Alaska's Prince William Sound shortly after midnight on March 24, 1989, causing the largest oil spill in U.S. history. A gleaming slick seeped across 100 square miles of the Sound before reaching shore. Winds gusting upward of 70 miles per hour delayed recovery efforts and worsened the onshore damage as fierce waves spat oil onto inland trees. When all was said and done, 11 million gallons of oil had spilled, desecrating 1,300 miles of coastline. Estimates of the number of birds killed were as high as 645,000. For many species of otter, fish, and birds, including bald eagles, it would take nearly two decades to recover.

Opposite: Oil from the Exxon Valdez *spill affected 1,300 miles of coastline.*

When the Coast Guard heard about the success Global was having along the Washington coast, they tagged along on a Global Diving response to see how the oil snare lines worked. They liked what they saw. "They used them quite a bit on the *Exxon Valdez*," Thom Davis said. "It certainly wasn't the answer to all their problems, but it was a viable resource for them."

In fact, an Exxon contact called Tim Beaver and told him the company was determined to buy "the world's supply" of pom-poms,

On Call and Ready to Respond

Following the *Exxon Valdez* spill and response, the oil transportation and oil spill cleanup industries alike were faced with a dilemma: how to have the capability to respond immediately to massive spills, in terms of both personnel and equipment, when there was not constant work to justify spending millions on new boats and new tools, and paying people to wait on standby for the next occurrence.

One solution was the creation of the Marine Spill Response Corporation (MSRC) in 1990. MSRC is a U.S. Coast Guard-certified not-for-profit that offers personnel and equipment to respond to oil spills, in an effort to clean up and minimize environmental damage. Global Diving has been one of MSRC's go-to contractors since the beginning and maintains a standby personnel roster of 140 to respond in emergencies.

"MSRC is a cornerstone contract for us," said Aaron Harrington, environmental manager at Global Diving. "It provides us a steady revenue stream and the ability to have a pool of guys that we can keep in the mix, engaged and trained."

Also formed in 1990 was Environmental Quality Management (EQM), an emergency response contractor that is frequently called into action on Environmental Protection Agency (EPA) emergency cleanups. Global is a teaming subcontractor with EQM and assists on responses in regions including the West Coast, the Southwest, Alaska, and Hawaii.

Global's MSRC and EQM participation has taken its crew members to events of international importance, including the responses following the Space Shuttle *Columbia* disaster in 2003 and the Deepwater Horizon offshore oil spill in 2010.

Global was contracted by the Washington State Maritime Cooperative (WSMC) to provide response services in 2013. WSMC provides oil-spill contingency plans and emergency response for vessels in Washington waters.

as the single manufacturer of the product couldn't keep up with the demand. Then-Vice President of Marketing Mike Craig had connections who knew of warehouses in Texas and New Jersey holding hundreds of boxes of pom-poms in storage. He and Beaver bought the pom-poms for Global, and in turn sold them to Exxon. They chartered a cargo jet to fly to Texas, then New Jersey, and then up to Valdez to deliver an entire planeload of the much-needed snares. In addition, Global's Seattle office worked overtime to provide significant amounts of materials to Exxon for the cleanup, including thousands of feet of oil booms, countless oil absorbents, and even skimmers imported from Denmark.

A crew of Global personnel made the trek north to join the cleanup crush that hit Valdez, Alaska, a town of 4,000 that saw its population more than double during the response efforts. Cab fares, hotel rates, boat charter prices all skyrocketed following the first wave of responders. Global Diving was one of the first companies subcontracted to help, not just with the oil snare lines, but also to pump oil from the tanker. Dave Usher and his company Marine Pollution Control (MPC) got the job to do the lightering of the vessel and called Global right away. "They were some of the first people aside from our own personnel that we called to work up there," Usher said.

He and MPC first became acquainted with the Global founders while they were with Crowley Environmental. Back then, Thom Davis was in Beaumont, Texas, working on the salvage of the *Marine Duvall*, which had sunk after a collision. He saw firsthand an MPC pumping system used to dewater the vessel. When he and the others started Global, they reached out to Usher to see if the pumping system could be available to them. Davis, along with Tim Beaver and John Graham, drove down to San Francisco to meet face-to-face.

Pete Shackleton (third from right) leads a team of MPC and Fred Devine Diving & Salvage crews aboard the stricken Exxon Valdez *in Prince William Sound, Alaska.*

"I looked into their eyes, saw their enthusiasm, and said to myself, these guys really want to make their mark in this business," Usher said. He then agreed to let Global be an agent for an MPC pumping system, and to take the unit to Seattle. The agreement was mutually beneficial: MPC had an agent it trusted in the Pacific Northwest, and also received a commission when Global used the system on its own projects; and Global not only had a state-of-the-art piece of equipment for salvage work but also gained exposure on jobs outside the Pacific Northwest when Usher called upon them.

MPC's subcontracting work took Global crews as far away as Antarctica, and now up to Alaska. "They were an integral part of the team we put together," Usher said. "I wanted to have Thom Davis and Tim Beaver up there because they were attuned to the environment and the weather, and I was confident in what they could do."

Usher's confidence in Global would prove to be warranted, and others on the response efforts would echo his praise of the company. J. H. "Mick" Leitz, a longtime Northwest salvor, was hired by Exxon as salvage master and worked closely with Global and MPC. "That was one helluva long, drawn-out affair," Leitz said of the *Exxon Valdez* response in the *Seattle Times*. "[Global] responded quickly, which is very important at a time like that, and they had good equipment. They did a very good job."

Global personnel were in Valdez for the next year and a half. It was worth it: the money made subcontracting "was our first real big chunk of cash," Jackie Lewis said. That year's Christmas party at Salty's, a restaurant on Seattle's Alki Beach, was one for the ages, with platters of crab and shrimp. Gary Duff's wife, Merilyn, even made him buy a suit for the occasion. "Global had arrived," Duff said. "It was a big deal. We were proud because we were successful."

Dave Usher of Marine Pollution Control "is one of the godfathers of the oil spill business," according to Thom Davis. Usher helped found both the Spill Control Association of America and the International Spill Control Organization. And to think, before his involvement in oil spill response, he ran a jazz record label with legendary trumpeter Dizzy Gillespie.

Since meeting face-to-face with the young Global founders in the late 1980s, he and MPC have had a strong, enduring relationship with Global Diving.

"I tell ya, I feel like their dad," Usher said. "I admire their enthusiasm and that they worked hard to move ahead. In that regard, their story is the American Dream."

As a tip of the hat to their old friend, Global Diving painted one of the anchors in front of the Seattle office red, to match MPC's logo of a red anchor.

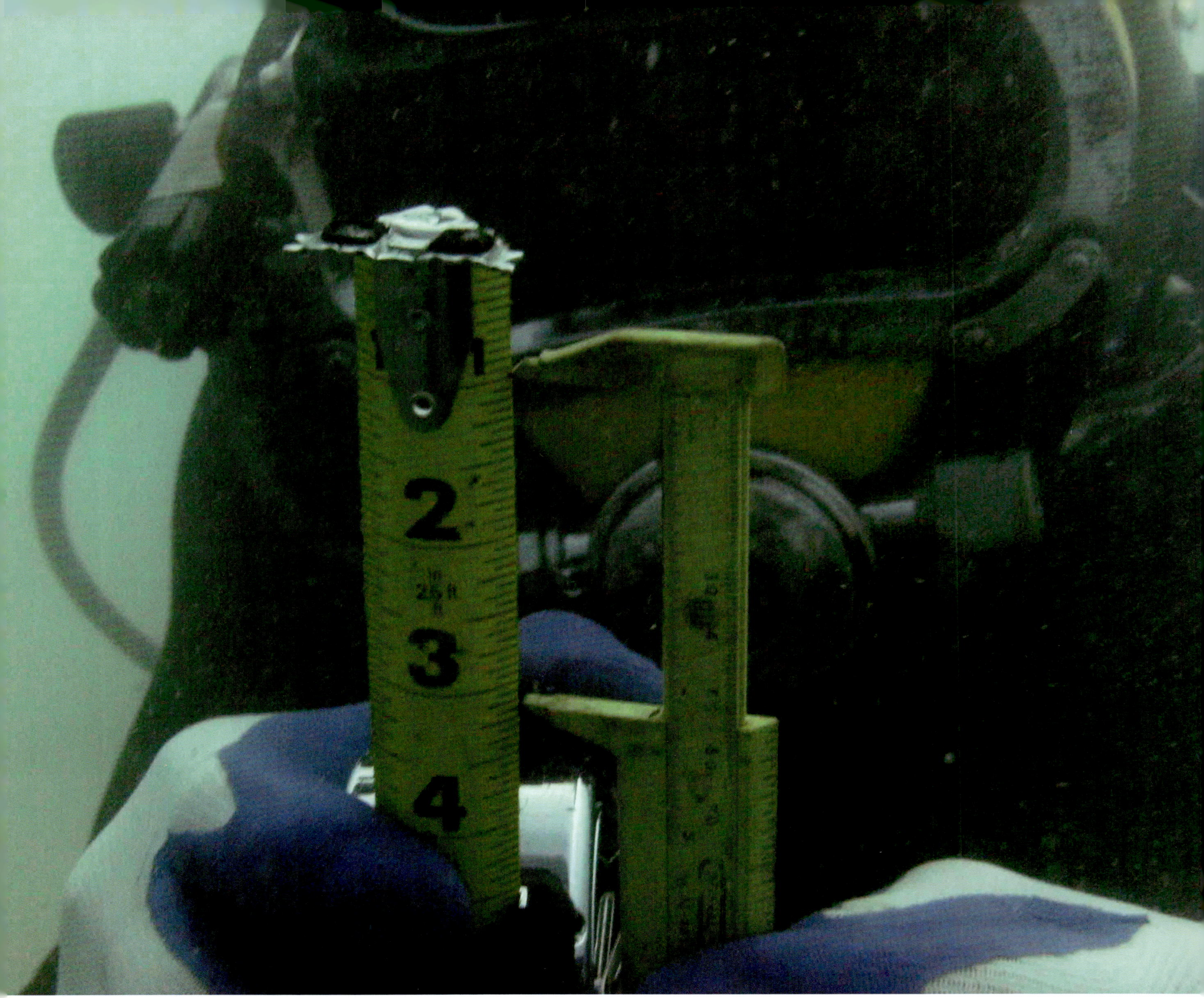
2
3
4

chapter 5

Moving Up to Go Deeper

With a fresh coat of paint, the tug *Spar* cruised at a casual speed on Opening Day of boating season 1990 through Seattle's Montlake Cut, with the University of Washington's Husky Stadium to its port and the Montlake drawbridge overhead. The *Spar* stood out, not for being the biggest boat, nor the most expensive, nor the most luxurious, but for being certainly the most spirited. A five-piece Dixieland jazz band hired by Gary Duff played on the upper deck as a consortium of carousers below, most several decades younger than the entertainment, swung and celebrated to the music, red Solo cups in hand. "We went down through the Cut and people went crazy," Duff recalled. "They were playing loud and everyone could hear it; it was an absolute hit."

After the successes of the late '80s, the Global Diving family was feeling good. However, one member of the family would soon be moving on, and an unexpected opportunity would arise to take a risky leap into a new diving niche.

Greg Harem Heads Back to School

Global Diving's first major change post-*Exxon Valdez* would be in leadership. The company's cofounder and first president, Greg Harem,

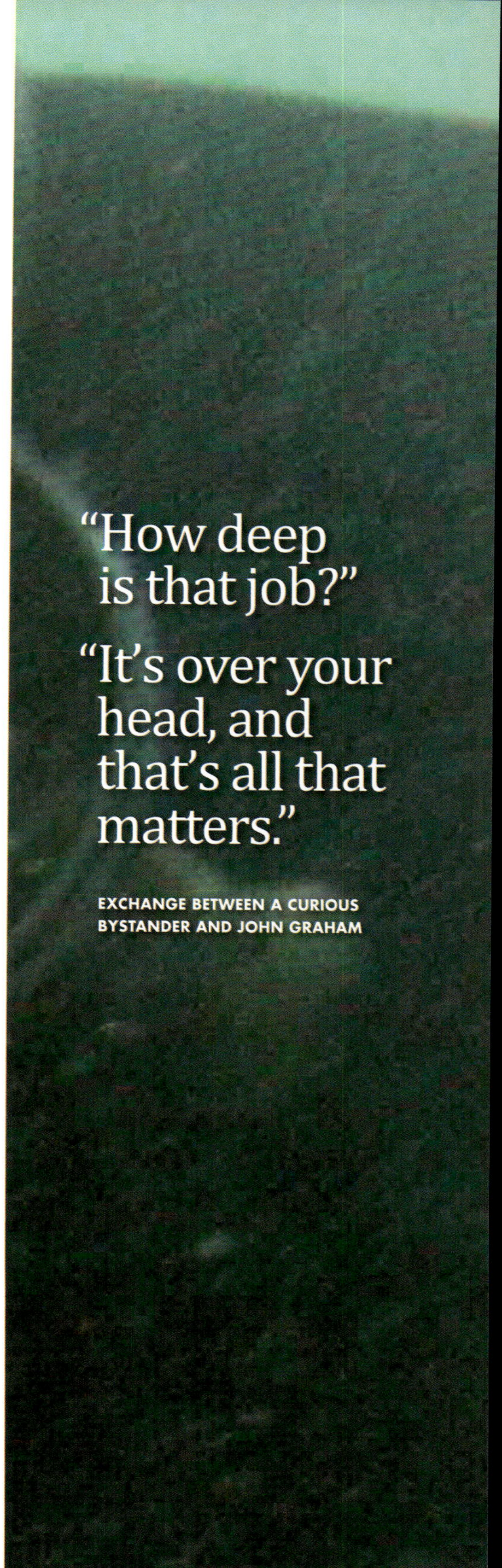

"How deep is that job?"

"It's over your head, and that's all that matters."

EXCHANGE BETWEEN A CURIOUS BYSTANDER AND JOHN GRAHAM

Precision and accuracy are acquired skills underwater.

The extended Global family aboard the Spike *and the* Spar *during the annual Tugboat Races in Olympia, Washington, 1991.*

decided to end his diving career and further his education. He left Global in 1991, after work on the *Exxon* cleanup was completed, and then began his studies at the University of Washington to earn a degree in engineering.

"Some people last a long time in the diving business, and some people don't," he said. "Your body makes the decision for you." Harem had experienced a case of "the bends" (decompression sickness caused by bubbles formed in the body) after a dive in which he strained a ligament, and then a couple cases of bruising caused by "skin bends." "I felt like it wouldn't have been smart for me to continue diving," he said.

Global lost not only its "compromise candidate" who could manage the always outsized, at times dysfunctional personalities of its divers, but also one of its great thinkers. "Harem was a genius; he might have been the smartest one of the bunch," Gary Duff said. "He could write good proposals and good bills, and would take a careful look at jobs to see what could go wrong and how that could be avoided."

After graduating from engineering school, Harem pursued his interests with an engineering firm that serves public entities including municipalities and special purpose districts, doing "one of my favorite things besides diving at Global": working with pumps and hydraulics, continuing in the vein of the salvage and construction work he enjoyed at Global, but out of the water.

Reflecting on his time at Global, Harem is most proud of how far the company came in a relatively short period of time. "Early on, I was in some ways embarrassed by our scrabbling around. Sometimes it was hard to take. Then the company began to take off and we gradually built a reputation for ourselves. We were working our tails off, and it was great to have the confidence of having made something out of nothing."

Top: Greg Harem during the Exxon Valdez *cleanup.*

Right: Global founders and co-owners (left to right) Tim Beaver, Thom Davis, and John Graham.

Most of all, though, he is proud of how hard the individuals in the company worked, Global's excellent safety record, and its reputation for finishing whatever it started. "People didn't hold back. They put everything they had into getting the job done, often with every bit of skills and stamina they could muster, and often in intimidating surroundings while under tremendous pressure.

"And yet, even while we were taking risks professionally, safety was always in the front of everyone's minds. You knew that someone always had your back, and as a company we took the time and dedicated the resources to make sure we did things the right way. I would expect that that aspect of the Global Diving culture has not changed."

Global added the 62-foot twin-engine Prudhoe Bay *to its fleet in 1992.*

Bell Bounce Bought

Tim Beaver became the next president of Global Diving & Salvage in 1991, and along with him came a tolerance for risk in business decisions heretofore unseen at the now 12-year-old company. Beaver was not content with the status quo of smaller-scale diving jobs and the occasional large-scale salvage or oil response project; he was intent on expanding the kind of projects and the size of projects the company could, and would, do.

One of the first acts of Beaver's tenure was to try to put to use equipment that Global Diving had bought but did not have work for. Before the *Exxon Valdez* response, Beaver received a call out of the blue from Anchorage, Alaska-based Underwater Construction. They were looking to sell a bell bounce system. A bell bounce allows for safer and more efficient diver recovery back to the surface on deep dives. The bell, a small chamber, is lowered to depth using a Launch and Recovery System (LARS). The diver climbs in, the bell is sealed,

Flush with cash after the Nestucca *job, Global decided to do some marketing. They designed and distributed an informational brochure at the behest of crewmember Mac McCarthy, who also staged and shot the above photo at Lockheed's Terminal 105, across the water from Harbor Island. "We were busy trying to prove to the world that we were substantial," Tim Beaver said. "We wanted to show that we're not just this little diving company with two hoses and a pickup truck; we do a lot more than people think."*

and decompression begins. The bell is brought on deck and then is mated to a decompression chamber. At the time, Global had no need for such a system, given the kind of work they were doing, and certainly not at the price being asked. "First of all, I didn't really know what a bell bounce was," Beaver said. "Then he said his company paid $375,000 for the system but would take $200,000. I said, 'I think you have a mistaken idea about us. First of all, what do we do with that kind of thing in Puget Sound? Second, we don't have anywhere near that kind of money.'"

Global's first bell bounce system would turn out to be an important step toward more efficient deep-diving methods.

Still, talks continued, especially after Global found itself flush with capital after the *Exxon Valdez* cleanup. Underwater Construction, hard up for cash and with no other potential buyers, eventually negotiated themselves down to a more reasonable figure. After consulting with John Graham, who decided the chamber would be a worthwhile addition at an estimated value of $25,000, Beaver made an offer. The two sides settled on $28,000 for the bell, LARS, and chamber. Global had little know-how and no work for the system, making an investment completely contrary to common business school teachings.

Deep Inland Work at Dworshak

"And there it sat for a while," Beaver said, until another fortuitous opportunity arose just as unexpectedly less than a year later. The U.S. Army Corps of Engineers put out the call for divers to attach a large copper plate to the face of Dworshak Dam, a gravity dam on the North Fork Clearwater River in north central Idaho. The fit had to be perfect, as a three-foot-diameter tunnel would be mined through the backside of the dam, breaking through inside the copper. The working depth was deep, at 225 feet below fresh water.

On the West Coast, there were several dive companies with the capacity to work at this depth but none for an extended period because of the large amounts of time lost to decompression, except,

The Air They Breathe

DIVING IS NOT AS SIMPLE AS CONNECTING TO A HOSE, breathing air, going under, and coming up when done. It is complicated by the uptake of nitrogen and the need to decompress when returning to the surface.

Divers, from sport to professional, use dive tables to determine the time needed to decompress based on dive time and depth. The deeper the dive, and the longer the duration at depth, the more time needed to decompress and decrease the nitrogen buildup in the bloodstream. If a diver resurfaces too quickly, the buildup can result in a potentially lethal internal formation of bubbles called "the bends." For professional divers on continuous gas working for hours at deep depths, the need for lengthier decompression times increases, as does the danger of the dive.

For decades divers breathed straight air (a natural mix of 19 percent oxygen and 81 percent nitrogen) while underwater, resulting in long periods of decompression due to the uptake of nitrogen gas into the bloodstream of the diver. The advent of "mixed gas," using other gases (especially helium, instead of nitrogen) to transport oxygen in the bloodstream, helped improve dive safety further. The use of helium eliminated the chance of nitrogen narcosis, also known as "the rapture of the deep," where divers experience mental and motor-skill impairment similar to that of having drunk too much alcohol. Lastly, the ability to custom mix the ratio of oxygen to inert gas avoided oxygen toxicity, which occurs when the partial pressure of oxygen gets too high.

Table 9-7. *No-Decompression Limits and Repetitive Group Designators for No-Decompression Air Dives.*

Depth (fsw)	No-Stop Limit	Repetitive Group Designation															
		A	B	C	D	E	F	G	H	I	J	K	L	M	N	O	Z
10	Unlimited	57	101	158	245	426	*										
15	Unlimited	36	60	88	121	163	217	297	449	*							
20	Unlimited	26	43	61	82	106	133	165	205	256	330	461	*				
25	595	20	33	47	62	78	97	117	140	166	198	236	285	354	469	595	
30	371	17	27	38	50	62	76	91	107	125	145	167	193	223	260	307	371
35	232	14	23	32	42	52	63	74	87	100	115	131	148	168	190	215	232
40	163	12	20	27	36	44	53	63	73	84	95	108	121	135	151	163	
45	125	11	17	24	31	39	46	55	63	72	82	92	102	114	125		
50	92	9	15	21	28	34	41	48	56	63	71	80	89	92			
55	74	8	14	19	25	31	37	43	50	56	63	71	74				
60	60	7	12	17	22	28	33	39	45	51	57	60					
70	48	6	10	14	19	23	28	32	37	42	47	48					
80	39	5	9	12	16	20	24	28	32	36	39						
90	30	4	7	11	14	17	21	24	28	30							
100	25	4	6	9	12	15	18	21	25								
110	20	3	6	8	11	14	16	19	20								
120	15	3	5	7	10	12	15										
130	10	2	4	6	9	10											
140	10	2	4	6	8	10											
150	5	2	3	5													
160	5		3	5													
170	5			4	5												
180	5			4	5												
190	5			3	5												

* Highest repetitive group that can be achieved at this depth regardless of bottom time.

Above left: A Deck Decompression Chamber, utilized to simulate in-water decompression, increased safety for divers.

Above: A page from the U.S. Navy dive tables, which recommend limits of decompression and no-decompression diving with standard air and mixed gasses.

GLOBAL
LTV Energy Products
Oil States Industries Division

serendipitously, Global Diving. With the bell bounce system that was collecting dust in their shop, Global could retrieve one diver in the bell, and while he was decompressing another diver could jump in, rather than having to wait hours for the first diver to slowly and carefully make his way up to the surface.

Global got the job, and set off due east for Idaho. John Graham and Bruce Humberstone headed the project. Global's supervisor ("diver driver") was Donald Moore, known to all as "the Shadow," a weathered veteran of the Gulf Coast who earned his nickname for his gray hair, ghostly pallor, and gaunt physique. They made a series of dives, all under the scrutiny of a Navy dive officer, and met the client's expectations enough to get a call to do more work the following year.

"That was a pivotal job because deep inland construction is a niche we've continued to steadily develop to this day," Beaver said. "We're the world's premier guys for that kind of work, and at the time we didn't have any way to envision what this project would lead to some 20 years later."

After using their wits to parlay their new equipment into a successful project with Dworshak Dam, Global had its eyes open to new opportunities but was careful not to turn away from its core competencies. They continued with shorter-term dive work and assisted on environmental responses, such as the cleanup of the *Tenyo Maru* tanker spill off the Washington coast in Neah Bay.

In addition, the company continued to mature. Marketing VP Mike Craig set up the company's first computers, with a Windows 3.0 personal computer at each desk. Now, all the inventory, sales tracking, jobs records, and other accounting that had been done by hand was maintained and shared digitally across a cable-connected network.

The Global family also continued to grow, welcoming diver Dan Pierson to the fold in 1992. The young diver was finally added to the call sheet after six months of "hounding" Beaver and Graham for

Top left: The Dworshak Dam team included, from left to right: Herb "Herbnoxious" Williams, John Graham, Steve "Four Buck" Eyler, Bruce Humberstone, and Scott Dimmick.

Top right: Mike Whetley stands with the cofferdam before it was installed on the upstream face of the dam.

Bottom: Dworshak is the third-tallest dam in the United States, at 717 feet.

work. "I remember when the call sheet was one page. There wasn't a second page, there wasn't a back side, it was one page," Pierson said. "It was a big deal to finally get your name on that and be part of the team."

He also remembers being impressed by the company's owners, who worked in the trenches (or at least in the water or on the beach) with their guys on all their jobs. "It was motivating. You weren't working for somebody that talked to the owners; you worked with the owners. The owners were right next to you, picking up oil off the beach."

Beaver recalls meeting Pierson for the first time. While he was working one day at the Harbor Island office, the Global headquarters' lone room darkened. Beaver looked up to find Pierson's six-foot, six-inch frame blocking the door. Beaver asked if he could help him, and Pierson responded, in his trademark straightforward manner, "I hear you are the best diving company in town, and I want to work for you guys."

"It was the shortest and best job interview Global ever did," Beaver said. The very next day, he took Pierson out with him on a small job in nearby Bremerton, Washington. Little did they both know that they would later work, and live, side by side in exceedingly close quarters, when Global decided to push the limits of the bell bounce system, making a play for a golden—but even more risky—opportunity.

Above: Crewmembers—including (front row, left to right) Ron Larsen, Mac McCarthy, and Clint "the helicopter pilot" Burke—celebrate a birthday during the Tenyo Maru *cleanup.*

Opposite: The Tenyo Maru *in better days. In 1991, the tanker collided with another vessel and sank, releasing 140,000 gallons of oil that washed up along coasts in Washington and Oregon.*

第三天洋丸
TENYO MARU NO3

chapter 6

Gold! Global Goes Treasure Hunting

In 1997, before the influx of dot-com and tech talent in the 2000s brought corporate campuses, trendy happy hours, and pricey lofts to Seattle's bustling South Lake Union neighborhood, the Cascade district, with its rows of old warehouses and small legion of streetwalkers, was the kind of place you might avoid at night, and would have little reason to be near during the day. Divers, by nature, go where others don't, and so business took Tim Beaver and John Graham to the Denny's restaurant on Mercer Street to discuss, of all things, treasure.

Specifically, $20 gold coins straight from the San Francisco mint that had not been seen above water since 1865. That year the SS *Brother Jonathan*, a side-wheel steamer, capsized off Crescent City, California, near the Oregon border, during a heavy storm on its way to stops in the Northwest territories. Two hundred and twenty-five passengers and crew drowned, including A. G. Henry, governor of Washington Territory. Also reportedly on board were over 300 barrels of whiskey, a menagerie of circus animals, and lots of money. The payroll for U.S. Army troops at nearby forts, treaty payments for tribes, and cash for federal banks were all still waiting at the bottom, some 285 feet below, for whoever could get to it first.

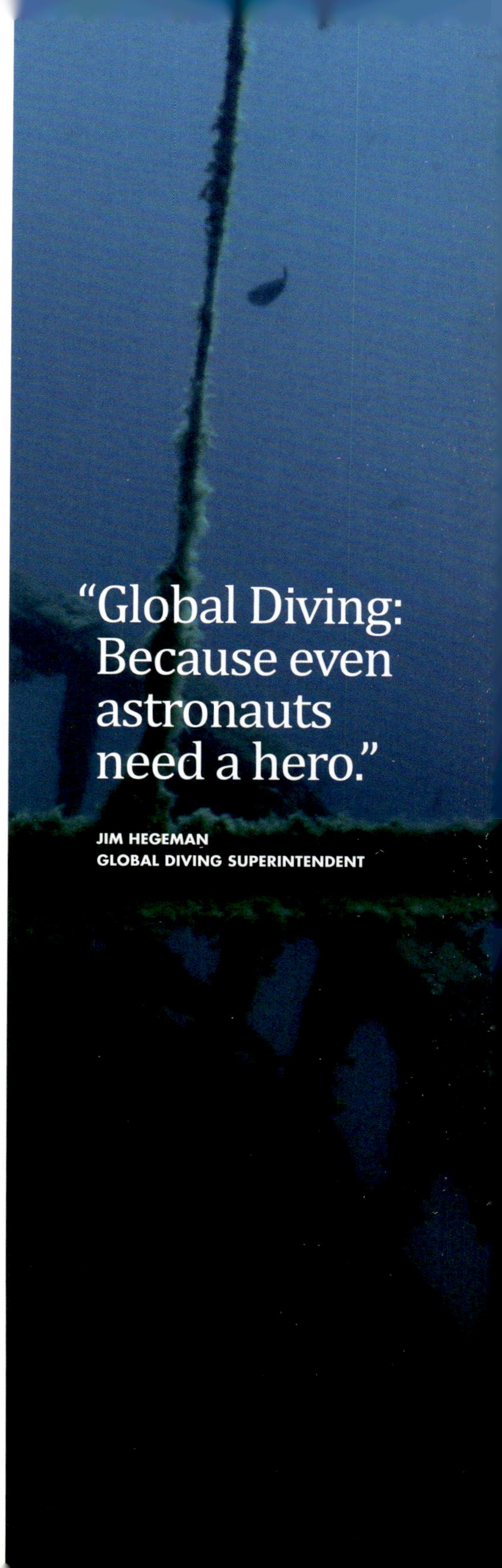

The $20 gold coins that Global divers retrieved from the wreck of the Brother Jonathan *fetched on average $5,000 each at a 1999 auction.*

Above: An 1851 oil painting of the Brother Jonathan.

Left: Pieces of the Brother Jonathan *washed ashore after its sinking. Here, Crescent City residents pose around the ship's wheel.*

Above: Harvey Harrington from Deep Sea Research.

Bottom right: This gold coin sold for $12,650 at auction.

Across the table at Denny's was Harvey Harrington, whose California-based company Deep Sea Research (DSR) had used sonar to finally locate the long-sought-after hull in 1993. After a prolonged $1 million legal battle with the State of California over the rights to the ship—and its cargo—that went all the way to the U.S. Supreme Court, DSR could proceed with exploring the wreck and retrieving its treasure. They agreed to give the state a share of the coins and whatever artifacts were collected. DSR then hired technical scuba divers to investigate the wreck, and they came back to the surface with coins and the report that there were many, many more lying below. Harrington had vetted Global Diving for the job, but didn't know how much it would cost, or how long the job would take. All he had with him at Denny's was a three-ring binder with pictures of the gold coins he knew were waiting in the rotting hull of the *Brother Jonathan*.

Accidental Treasure Hunters

For Beaver and Graham, who were later joined at Denny's by Thom Davis, the prospect of treasure hunting was not particularly appetizing. Harrington wanted them to work for a percentage of the project. "At that point we told him, you have a misunderstanding of who we are," Beaver said. "We can't afford to do that. They thought we were this big company."

The Global Diving heads came up with a day-rate figure, and DSR balked. The back-and-forth between the two sides continued, but a commitment was difficult to confirm. The deal was only closed after Beaver and Graham caught a red-eye to New Orleans, rented a black Lincoln Town Car, and drove across the Gulf Coast to the Florida panhandle, where Harrington lived, to meet face-to-face and nail down the particulars. With three hours of sleep they met Harrington for breakfast and after a long day closed the deal, had dinner and drinks, and then hopped back into the Lincoln and drove all the way back across Alabama, Mississippi, and into the Big Easy to catch their red-eye flight home.

Stepping into Sat

The *Brother Jonathan* was a leap of faith for Global Diving, and a crucial first in the company's history. It would be Global's first foray into saturation diving, or simply, sat (see sidebar). "You can't grow unless you leave your comfort zone," Tim Beaver said. This would be a maxim that Global has repeatedly maintained throughout its history.

Up until the *Brother Jonathan*, the vast majority of sat diving in the United States had taken place in the Gulf of Mexico for offshore oil support work. Nevertheless, Global Diving had purchased a bell bounce chamber from Underwater Construction in Anchorage, and had already used the equipment for the Dworshak Dam project. They had experienced sat divers on their team, including Bruce Humberstone, who cut his teeth as a diver in sat while working in the Gulf. Humberstone had left his full-time position at Global a few years earlier to work freelance, and had been brought back for this project to lend his sat expertise. He and good friend Herb Williams retrofitted the tiny chamber from the bell bounce system to become a mini-sat system, and both lived together in the chamber until the job was done.

Committing in Crescent City

With their first saturation system, later dubbed Sat I, readied for the *Brother Jonathan*, a Global caravan made the nine-hour drive down to Northern California. Dave Partlow was dive supervisor, and Dan Pierson was deck support, excited to be involved in his first sat job. Humberstone and Williams were the divers. Squeezed between owners Tim Beaver and John Graham in Graham's pickup truck was David DeVilbiss, a young hand who began part-time work for Global while studying at DIT. He started out painting the Global landing craft *Prudhoe Bay* but then showed promise as a diver, welder, electrician, and salvor, and so was asked to join the *Brother Jonathan* on deck as a tender. DeVilbiss took a leave of absence from DIT, the first of

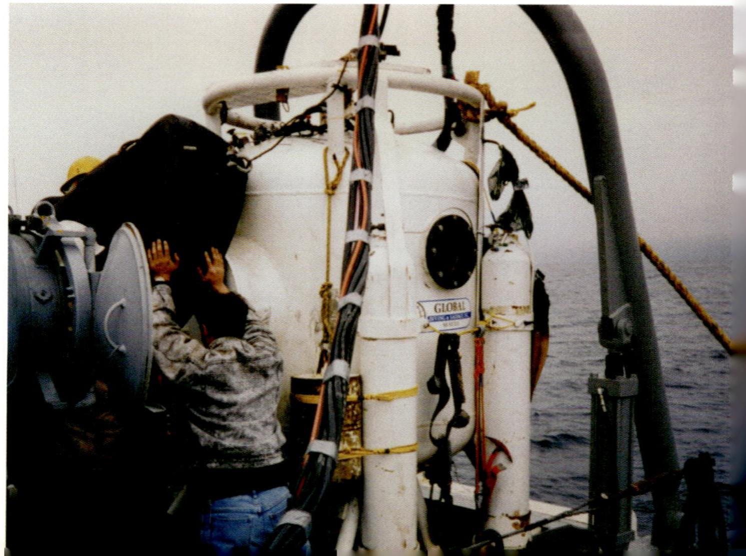

Top to bottom: A look inside the living chamber during the early days of Sat I; the living chamber rests on deck; Global personnel mate the bell to the living chamber.

Below: Lowering the bell with Global's trademark protective tire over the trunking.

Right: All a diver could need in sat: paper towels, hot sauce, and athlete's foot spray.

"Sat Is Where It's At"

For most divers, saturation diving, or simply sat, is the pinnacle of their dive career. The projects are lucrative, often with the highest stakes for company and client, and are the ultimate test of skill and determination, with wet pay to match—hence, the divers' saying "sat is where it's at"—but it's certainly not for every diver, especially the claustrophobic.

Saturation diving is used on deep projects lasting days, weeks, or even months. To save time and greatly lessen the risk of decompression sickness ("the bends"), divers are "saturated" at pressure in a system of chambers where they will live for the duration of the job, only decompressing to surface pressure after the work is completed or their rotation is up. The divers sleep, eat, and spend all non-working time in an on-deck living chamber. They are transported to job depth by a smaller "bell," which mates to the living chamber and is raised and lowered by a Launch and Recovery System (LARS). The bell is lowered to the project depth, then the divers swim out and work (already pressurized to that depth), return to the bell when their shift is done, and are brought back up to the surface, where the bell mates with the living chamber, while the pressure inside the chamber is maintained. Food, laundry, and other effects are provided on board the ship, then given to the divers through air locks. A large support team attends to the divers' needs and monitors the oxygen levels in the bell and deck chamber around the clock. The two to six divers in the chamber live, eat, and sleep together for the duration of the project, crammed into a small space they cannot leave except to crawl into the tiny bell that will take them to work. Due to a helium-oxygen breathing mix that can be 97.5 percent helium, they sound like Donald Duck. During downtime divers may read, listen to music, or practice free throws on a Nerf basketball hoop. Regardless of what's on the menu, the divers douse their food in hot sauce to combat the air pressure, which desensitizes the taste buds. Though they eat well, many divers lose weight from the calories burned in the water and exit the saturation diving complex skinnier than they entered it.

Top left: David DeVilbiss, at left, and Scott Dimmick.
Top right: The American Salvor.
Bottom right: Cruising past the Golden Gate Bridge.
Bottom left: "Old school super tender" Ron Larsen.

many (DeVilbiss claims his was the longest time needed to graduate from the program, but it is a distinction also claimed by several other Global personnel). On board the *American Salvor* (ex-ferryboat *Manatee*, ex-*Arctic Salvor*, and now Crowley Maritime's flagship salvage vessel), chartered for the job, Superintendent Dave Partlow was surprised to see the fresh face bringing his dive gear aboard. Upon learning that young DeVilbiss was still in school, Partlow put him in his place: "Hell, boy, you're not even a tender. You're a tenderette!"

Just as DeVilbiss was deemed not quite ready for prime time, the set-up on board the *Salvor* was similarly makeshift, more like "sat-lite." The operation was run on a shoestring with most equipment borrowed or rented, including the control van and an environmental control unit on loan from DIT and rebuilt. Sat chambers are always cramped, but this was something else: a 54-inch deck chamber designed for one person to do surface decompression was now housing two six-foot-plus men, their gear, a Coleman camping porta-potty, and a shower made from a complex assembly of pumps, valves, and regulators. The MacGyver-like shower contraption became known at the job site as the "Humminator" hot water delivery system, named for its crafty inventor, Bruce Humberstone. It was all part of the job for Humberstone and Williams, who had been roommates in the Gulf of Mexico in the early '80s while starting out in the industry. "When you're in sat, you're always in a small area," Humberstone said. "That was just a *little* bit smaller than normal."

It was to be Humberstone's last time in sat, and once he and Williams were put to pressure, or "committed," the team on board heard nary a grievance. "Those guys had to be miserable in that thing, but there weren't any complaints," Dan Pierson said. "Herb and Bruce jumped into that thing headfirst. That's a testament to Global and the people who work here."

Top to bottom: A view from below of sea anemone beating Global to the buried treasure; crew lower crates for collecting items found; artifacts from the Brother Jonathan*. The Del Norte County Museum has a collection dedicated to the shipwreck, California's worst maritime disaster.*

Left to right: Dan Pierson suits up; Dennis Tsugawa hard at work; Tim Beaver, excited for sat.

Over the next 18 days, they explored the wreck and collected gold and other artifacts. The Global Diving team above surface gained valuable experience on how to run a successful sat operation, a skillset that ended up being worth more than gold.

Round Two and an Important First

Deep Sea Research and the State of California were so pleased with the findings during the operation that they hired Global to come back in 2000 to collect more materials off the bottom. This time, Global brought down an improved Sat I with a larger living chamber, rented from Professional Divers of New Orleans. "We upgraded to 1970s technology at that time," said Dan Pierson, jokingly. He was put under pressure in sat for the first time and enjoyed the experience.

"I enjoyed the long bell runs and being able to dive for five or six hours at a time. That's my perfect place—just being in the water and working. You don't have any other distractions around you. To me, that's the most peaceful, easy spot to be: on bottom, working, talking to one person, with one goal ahead of you."

When another one of the divers in sat was unable to complete the work, Beaver jumped at the chance to dive sat for the first time in his career. But before he could, he had to make sure Pierson was agreeable to working with him. "The operations team called down to the chamber and asked if it was okay if Tim came down and dove with me," Pierson recalled. "I was thinking, Why are you asking me? He owns the company! But that's the respect that this company has for its divers." Beaver stayed in sat for 23 days, including decompression, and kept a log of his only experience saturation diving (see sidebar, page 84).

The goal was to find gold and bring it back, which Beaver, Humberstone, Pierson, and Williams did in spades. Over the two missions, 1,207 coins were retrieved. One can be found today at the historical museum in Crescent City. After the first round of dives, DSR made $5.3 million at a 1999 auction from the sale of 1,006 coins, about $5,000 a pop. One flawed coin minted with an inverted date went for $115,000 on its own.

Though the Global Diving owners never planned to be treasure hunters, the *Brother Jonathan* would prove to be a critical milestone for future saturation diving jobs, which would become more lucrative, more complicated, and more commonplace for the company.

An advertisement for the Brother Jonathan *auction catalog by Bowers and Merena.*

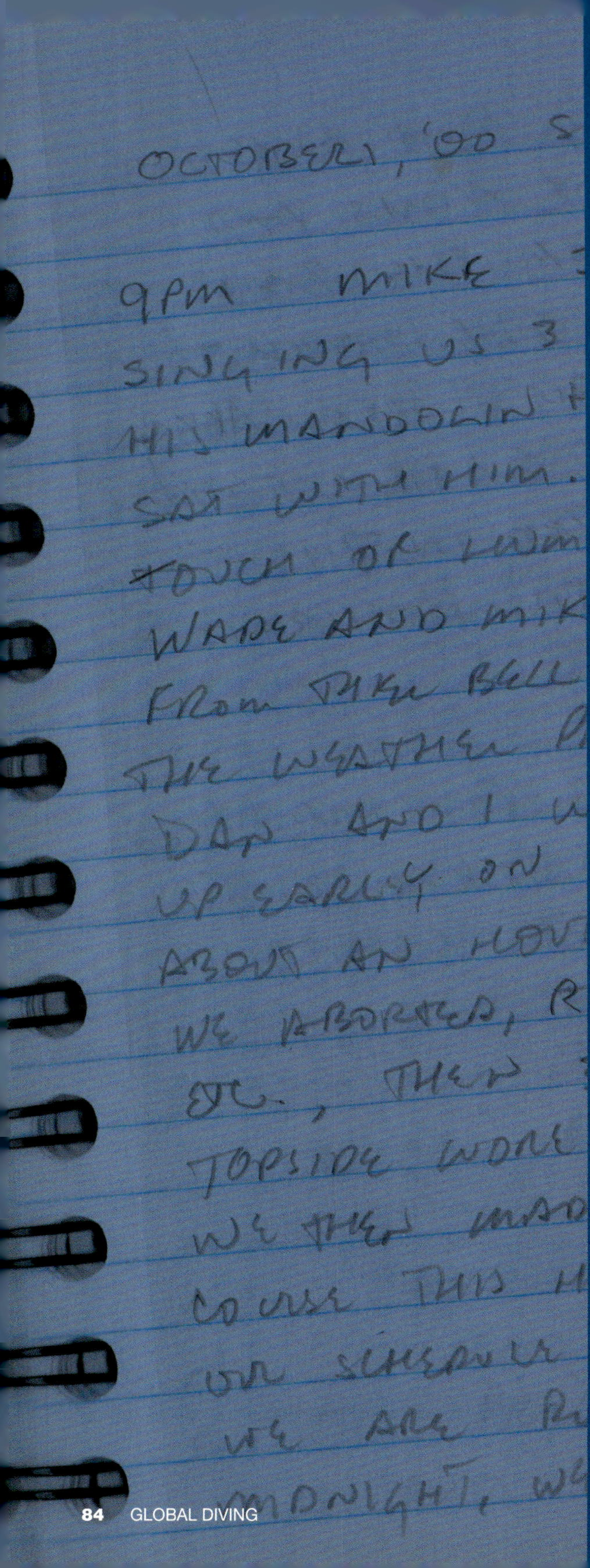

Notes from the Bell

While saturation diving on the *Brother Jonathan*, Tim Beaver kept a log during his 23 days in sat. The following are excerpts from his log.

September 22, 2000, 1845:

"My bell partner is Dan Pierson, and he is kindly showing me the ropes. Dan and I are to make the first bell run, but haven't decided who will make first dive. Called home twice and got machine both times. Left message on second call but didn't mention sat."

September 23, 2000, 1545:

"Slept from 0830 until 1500. Bell run was more work than anticipated, but survival was achieved. There is a specific sequence to everything 'in here' due to the confined space, and the extreme pressure differential. We are living under about 100 PSI, so every article of clothing, tool, meal, etc. must be locked in and out."

"Each diver dives for four and a half hours. It is about a toss-up for which is better/worse: sitting for five hours in the bell, or scrabbling about on the bottom. The act of riding up and down, and of getting the diver in and out of the bell is a performance involving the skills of the contortionist and the actions of a good friend. It works, but just barely. Danny is six-foot-six and leaves little room for my sorry ass, even though he tries, and it couldn't be done at all without his help and knowledge."

October 3, 2000:

"Still standing by waiting on the weather. It has become increasingly windy, with commensurate chop and swell building. The ship shudders with each swell and pitches and rolls with occasional violence. Wrote some letters, made up a calendar, formulated plans. Reading T. S. Eliot, the pompous S.O.B. Little else going on."

October 6, 2000:

"Good bell run last night. This noon to midnight shift is agreeable. Bruce gave us the news that we were set to pull anchors on the 9th at midnight. This helped morale tremendously as we are all ready to get out of here. By the time we are decompressed, it will be 23 days for me and 40 for Dan."

October 14, 2000:

"Broke the seal and awkwardly stumbled forth. Strangeness of 'normal' voice. We stumbled around and eventually went to bed. Steaming towards Long Beach Harbor Pier 49 – Crowley's dock."

The American Salvor *experiences rough seas. Inclement weather caused numerous delays for the 2000 mission, leading to more time in sat for the divers, including Tim Beaver, who detailed the experience in his personal log (at left).*

DUI

chapter 7

Salvaging Tough Situations

Beginning in 1996 and continuing through the next year, Global Diving embarked on a project much less glamorous than treasure hunting. Instead of gold, they were knee-deep in toxic chemicals in Eagle Harbor, west across Puget Sound from Seattle on Bainbridge Island, cleaning up the remnants of a collapsed industry.

A creosote plant had been in operation in the harbor since 1904 and provided work for generations of Bainbridge residents, treating timber with the wood preservative and shipping the longer-lasting product around the world. The chemical was considered a godsend for the small town, which even named itself Creosote. Bought by the Wyckoff Company in 1964, the plant operated until the late 1980s, and was officially listed as a Superfund Site in 1987. Now better understood, creosote is highly toxic to marine life and harmful to humans as well, something Dan Pierson experienced firsthand while working on the Eagle Harbor environmental remediation project.

"Creosote is extremely corrosive," he said. "You get it on your skin, and it's like having baby oil on you on a 110-degree day under the sun." This time Global Diving was hired for the job as a prime contractor—not answering to Manson Construction, or Foss Environmental, or another partner, but straight to the Department

"How's the weather up there?"

DIVER CODE FOR "IS THE CUSTOMER LISTENING?"

From left, Global crewmembers Travis "Spot" Havens, Jim "60-Hour Monkey" Givan, and Mike Szymczak assist in the salvage of a mangled semi below the Sam Hill Memorial Bridge, on the Columbia River between Oregon and Washington.

of Ecology, the Environmental Protection Agency, the State of Washington, and a handful of other government and regulatory agencies overseeing the project.

Global got the job because of the company's experience in a variety of disciplines—site demolition, diving, marine construction, and environmental response work—and its ability to incorporate all these talents on the same project. One new requirement they faced as the prime contractor was the stipulation for a lengthy, site-specific safety and health plan, which Tim Beaver (who was an English major after all) decided to write himself. The company looked at outsourcing the writing of the plans, but when the price tag came back at 10 percent of the total value of the project, Beaver successfully worked through it in-house, saving nearly $30,000.

Divers could see their reflections in the creosote pools on the bottom of Eagle Harbor.

The work on the remediation involved removing an old wharf. Crews cut pilings topside and underwater and completed other underwater demolition work, then had to carefully remove the pilings, which were required to be cut two feet below the mudline. They also removed pools of raw creosote lurking below sand and mud. Divers in Tyvek hazmat suits went through full decontamination to get in and out of the water. Global's work went smoothly on the remediation (a process that was still ongoing as of 2013, long after Global's involvement on the shoreline project was complete), which was yet another valuable learning experience for a company still evolving in its second decade.

"Eagle Harbor gave us an understanding of what regulatory agencies desire and how to manage their expectations," said Devon Grennan, who was brought on by Thom Davis in 1995 as an environmental technician and became president of Global Diving in 2009. Diving as a subcontractor was one thing. Priming

The creosote plant at Eagle Harbor was a crucial economic engine on Bainbridge Island for much of the 20th century.

projects—answering directly to a client, devising a plan, selling it, and completing it while still making a profit, with little in the way of a safety net—brought Global into the big leagues.

"Actually Doing Something"

Devon Grennan's first job at Global would have made most people reconsider their new place of employment. But standing in the pouring rain, well past midnight on a balmy August night, setting and picking up anchors for preventative booms in advance of Seattle's annual Seafair festivities—"I thought it was cool," Grennan said.

After graduating from the University of Washington with a degree in political science, Grennan was waiting tables at Panda's Chinese Restaurant ("because you don't do much with a political science degree") until his father, Pat, put him in touch with Thom Davis, an old colleague from their days together at Crowley Maritime. The younger Grennan had always been fascinated by marine work, his family's chosen trade, and even wrote his dissertation on the effects of the Oil Pollution Act of 1990 (OPA '90). "I thought it would be exciting to go clean up oil spills as a career," he said.

"I had friends in law school and others working for Amazon, and here I was in the middle of the night on the water. One thing that attracted me to Global is, you're actually doing something. You can point to something we're building, or something we clean up, or salvage, or make better, as opposed to 'I transitioned numbers in a computer program.' You can't lay your hands on that."

Grennan joined the still-small Global staff, becoming one of six people who regularly worked in the Harbor Island office: he and Davis, the environmental division, shared the upstairs with Greg Fladseth, the remediation manager who helped bring the Eagle Harbor project to Global; Tim Beaver and John Graham were downstairs; and Jackie Lewis kept everyone in line and everything else running smoothly. He worked closely with Davis, absorbing his industry know-how and his way with clients. "Thom was like a second father to me for four years of my life," Grennan said. "He took good

Devon Grennan comes up for air on the 520 bridge cable replacement project.

Global's Out-of-the-Water Partners

OVER THE LAST 35 YEARS, GLOBAL DIVING HAS FORMED STRONG RELATIONSHIPS with its vendors, suppliers, contractors, and other in-the-water partners to help get the job done on the company's dives. Just as important, but not as visible to those outside the company, are the enduring partnerships Global has forged with its partners in legal representation, finance, and banking, chief among them James Bauer, Loren Hostek, and Clarence Banks.

Bauer's firm, Bauer Moynihan Johnson, specializing in maritime, fisheries, and transportation law, had in earlier years represented Crowley Maritime Services, which included Crowley Environmental. It was there that he first met some of the initial Global founders. "They were young guys who worked hard, did a good job, and took care of their customers," he said.

Shortly after incorporating, Global hired Bauer for counsel. Over the next three decades he provided general legal services to the company and watched it grow and develop. Many of the company's customers were concurrently clients of Bauer, so he saw the company from both internal and external perspectives. Through it all, he credits what he saw from the young Crowley divers with allowing the organization to grow and expand to the level it has.

"There was little marketing or advertising," he said. "They developed the company year after year, incrementally, by doing a good job, taking care of and being well-received by their customers, and offering additional services as opportunities arose."

Hostek, a certified public accountant (CPA), began working with Global in 1988 to help the fledgling company develop basic accounting systems. Hostek's experience working with clients in the fishing industry gave him an understanding of the financially volatile nature of diving, something he's helped the company manage over the past 25 years. He's seen Global's growth firsthand.

"It's been a privilege working with Global," Hostek said. "I've watched a company go from just making wages every year, and if they were lucky getting a bonus to take home a new truck, to one that has an international reputation for the quality of the work that they do."

Clarence Banks became Global's loan officer at Bank of America in 1993, after the company had finished work on the *Exxon Valdez* response. Tim Beaver, Thom Davis, and John Graham brought him up to speed on the ins and outs of the diving industry, but it was then-office manager Jackie Lewis who helped him understand Global's financials. "She was amazing back then, and she's amazing today," Banks said.

Banks has helped make projects possible by facilitating financing, which allows Global to pay its workers and buy or rent equipment before they've been paid by their clients. Banks was also instrumental in financing the purchase of Global's offices on West Marginal Way.

After Banks moved to HomeStreet Bank, Global came with him. "Every time I hear 'Global Diving' or a report of what they've done lately, it fills me with more pride than you can imagine," he said. "The only thing I have more pride in is my kids. If my career ended today, I could survive on just the positive memories I have of working with Global and what they've done over the years. I was fortunate enough to be along for the ride."

care of his clients and set a good example of workmanship in the field and doing it right the first time. Thom was cognizant of saving money for his clients and trying to minimize costs as much as possible. On one level it wasn't a savvy business move, but it was the right thing to do rather than take advantage of the situation."

Moving to West Marginal Way

The Global Diving offices on Harbor Island were temporary barracks that had been inhabited for the previous 30-odd years, since being vacated by the military after World War II, by various businesses including Global-precursor Marine Oil Pickup Services. In 1997, the Port of Seattle finally decided to redevelop Harbor Island and do away with the long-interim facilities, including the office that Global had rented since day one. By this time Global was also renting a yard a block down on 13th Avenue Southwest, as well as part of a warehouse around the corner and another chunk of warehouse next door to the main offices at 2763 13th Avenue Southwest. It was high time to find a new home.

With generous help from the Port of Seattle, the company found a larger facility on nearby West Marginal Way, in the shadow of the West Seattle Bridge. The building was constructed in 1931 as the corporate headquarters for General Construction and enlarged 10 years later. It represented a huge improvement over the Harbor Island office, in terms of both the condition of the office and functionality. On Harbor Island, Global was cramped in the main office. The new office allowed management and support staff to have individual offices on the main floor, and the ground floor would be the divers' territory. They would also be able to add more buildings nearby, and over the next decade would themselves build a warehouse just north of the main office and purchase two yards and a large fabrication facility with more storage space across the street.

Even more crucial, Global was no longer renting—the owners were able to buy their own facilities with the help of Bank of America and longtime loan officer Clarence Banks. "Having a new facility was a great image boost, and it made Global more efficient and more functional," Banks said. "They had more space to be a cohesive unit, and they had room to grow as an organization."

Beaver, Davis, and Graham were also able to outfit the office the way they had always wanted. In the main entrance stood a "Jake," the universal name for a stuffed dive suit. Davis had purchased the helmet for their Jake during the 1986 *ARCO Anchorage* job, from a Port Angeles widow whose husband had been an old-school diver. The diving dress came from Al Jett of General Construction, the weight belt and brass boots from elsewhere. They also began filling their walls with photos of boats, past jobs, and anything else related to the water. Their decorations were in part homage to Mark Freeman, of Seattle's Fremont Tug. "When you went to Mark to rent a tug or just say hi, you walked into an office that had all this history," Beaver said. "As a young guy, I used to think that was the way it ought to be. A business has to have a little bit of soul in it, as well as a nod to history."

After settling into their new space, Global was ready to make company history again, this time as a prime contractor for a major marine construction project.

Cable Adjustments

By the late 1990s, Global had become more aggressive in bidding on larger marine construction projects, driven by equal parts necessity and ambition.

The "gravy" of oil spill response was harder to come by, in large part because of the subject of Devon Grennan's senior dissertation, OPA '90. The Congressional Act, passed in response to the *Exxon Valdez* disaster, mandated that companies plan for

Clockwise from top left:

2006 Seahawks NFC Championship tailgating.

Visitors to Global's Seattle office are greeted by the company "Jake."

Left to right: Lori "Luscious" Baron, Randy Baron, Tim Beaver, Nerissa Beaver, Kim Immel, and Frank Immel.

At a Global shindig, left to right: Katie Cantrell, Shawntel Johnson, Trinity Ng-Yeung, and Brooke Boge.

oil spill prevention, whereas to a large extent before, they were just a cost of doing business. As a result, there were fewer oil spills, which was a damn good thing, but it meant Global had to fight against larger environmental operations for less response work.

Additionally, Global management was confident they could successfully helm the bigger, expensive projects that were being won by large competitors. "We drank the Kool-Aid that we were the best diving company to work for," Grennan said. "We were small and crafty, but there were bigger diving companies out there that were eating our lunch on the big work. I kept thinking, We can do that stuff, can't we?"

Grennan certainly wasn't the only one at Global thinking along those lines, but it was not until 1999 and a project for the Washington State Department of Transportation (WSDOT) that Global was finally able to prime some "big work" when it came to marine construction. Grennan assumed a project-management role on a cable-replacement project for the SR-520 floating bridge, connecting Seattle across Lake Washington to the city of Bellevue. The 520 bridge needed routine maintenance on its anchor cables, which involved swapping out beat-up two-and-three-quarter-inch structural strand cables for new ones. This had to be done in a rotation so as not to leave the bridge ends unsecured. Global's plan, originally drawn out on a napkin by Tim Beaver and David DeVilbiss, would use the *Prudhoe Bay* and a rented wire spooling machine, which (to the dismay of the renter) would undergo some Global modifications.

More than 190,000 people cross the SR-520 floating bridge every day.

Like most of the larger projects Global worked on at the time, this was an "all-hands" effort, bringing the whole crew out of the office and on-site to assist. Beaver, who as part of a WSDOT requirement had to become certified in erosion and soil storm water management ("On a freaking floating bridge job!"), ran the overall job on the water. However, it was Grennan who was volunteered to oversee the difficult and strenuous work done inside the pontoons, along with Aaron Harrington. The personnel inside the pontoons had to de-tension

Top left: David DeVilbiss.

Top right: Devon Grennan holds a cable while preparing it for socketing inside a pontoon.

Bottom left: A crewmember guides the lowering of a cable along with a crossbar full of equipment needed by the diver.

Bottom right: The rental spooler that underwent "some Global modifications."

Top: The New Carissa *after it had run aground.*

Bottom: Global crews, with Thom Davis on the phone, aboard the beached vessel.

opposing cables while at the same time using immense jacks, after removing tiny but stubborn pins that held the wire in place. The crews inside persevered through conditions that were "hotter than the hubs of hell," according to Harrington. "It was so hot and humid that we had to push in air all the time. You were drained by the time you got out of there because you were sweating all day."

The cables were brought up using a topside spooler. After the new cable was fed into the pontoon, it had to be cleaned with an aromatic solvent. "We had fans going, but we were thinking, We're all going to pass out and die inside this giant concrete box and they'll find the bodies down here," Beaver joked.

Once the pins were removed, that cable had to be replaced, which made for some long days, as the pontoon crews started their day before the divers and were still working long after they had gone home. Still, the team completed the project "by the skin of our teeth," Grennan said. "I felt like I was in over my head, but it was awesome."

"For a diving company to pull off that kind of project without using a large crane was pretty unheard of," Beaver added.

New Carissa

Global would again attempt to pull off the improbable on a subsequent salvage job, but this time they were unable to turn a bad situation around.

The *New Carissa*, a 639-foot Japanese-owned cargo ship, ran aground north of Coos Bay, Oregon, following a severe winter storm on February 4, 1999. *The Oregonian* eventually dubbed her "the ship that wouldn't go away, or even die," due to a salvage and wreck-removal effort that went awry and left her on the beach much longer than originally expected. A week after the running aground, the Navy and Coast Guard command team attempted an in situ burning of the 400,000 gallons of oil on board. This only set the ship on fire and split her in two, the stern section from the bow. After pumping fuel from the cargo hold, the plan was to tow the bow 248 miles to deep water and sink it, but gale-force winds and high seas caused weeks of delays.

Point Wells spill, Indianola, Washington. An example of the appreciation local communities have when Global visits.

When they finally did try to tow it out, the line snapped, setting the *New Carissa* adrift and grounding her anew. Each miscue only heightened the already high media interest in the *New Carissa*.

Any hopes of salvage went crashing away with the bow. When Global joined the wreck removal and oil response efforts, "it was a big, chaotic mess," said John Graham. Global had recently bought a one-quarter interest in Portland-based Fred Devine Diving & Salvage, who formed a joint venture with a New Jersey outfit to remove the stern section, and so Global became involved through Fred Devine and Mick Leitz, the famous Salvage Master. Global and Fred Devine crews worked on the *New Carissa* for nine months, not leaving the site until November 1999, exhausted and ultimately unsuccessful.

From left, Ryan "The Ripper" Grennan, Sammy Reppman, Thom Davis, and Devon Grennan get their hands (and more) dirty inside the New Carissa*. Due to ongoing bureaucratic feuding, the wreck of the* New Carissa *was not completely removed from Coos Bay until nearly a decade after Global crews had left the site.*

"It was an excruciatingly hard, physical job," Thom Davis said. "It wasn't a total loss because we gained so much knowledge, cleaned up a huge amount of oil, and salvaged a lot of steel off that thing, but it was the first time we'd ever been involved in a big, unsuccessful operation, and it was heartbreaking."

"Had we got this off, it would be an epic salvage tale to end all tales," Tim Beaver said. "Because we failed, it's not."

Bill Milwee, an independent salvor with experience on over 150 shipwrecks worldwide who headed the salvage efforts, was asked by *The Oregonian* if he would have done anything differently on the job. "Yes," he said. "I wouldn't have answered the phone."

The bow was eventually towed out, torpedoed by the Navy, sunk, and buried at sea, but it was not until 2008—and a series of nasty legal battles—that the stern and the rest of the wreck was removed by Titan Salvage. The *New Carissa* changed how the Coast Guard managed its incident command. It also sent weary and crestfallen Global Diving & Salvage to do some soul-searching.

chapter 8

Safety, Quality of Service, and Growth: Branding the Way

"A lot of things had changed after being gone for nine months," Thom Davis said of the time following the taxing and ill-fated *New Carissa* job. "I was tired, burned out." Things were changing at Global, and were about to change even more in the near future.

While Davis and others labored in Coos Bay, Oregon, Mike Langen led a team at Rocky Reach Dam in eastern Washington State. This was the first major, long-term inland marine construction project for Global, working as the diving subcontractor for Dix Corporation, yet another company milestone and successful venture into a new arena. Langen, who only months earlier had been recruited as project manager to fulfill Global's vision to enter the inland construction niche, bid the project as a turnkey underwater construction package. He pulled an all-nighter on the proposal, building "a better mousetrap" to win the job. David DeVilbiss and Dan Pierson were the lead divers, working on the dam in midwinter and at night (so as not to stop generating power at peak hours and to avoid the summer salmon runs) installing a prototype fish bypass channel on the upstream side of turbines to direct salmon fingerlings around the turbines. Rocky Reach was a lump-sum job that would keep Global busy for weeks at a time, a marked and welcome change from day-rate projects with a short time frame.

The job below often needs to be done regardless of the weather above.

Above: Rocky Reach Dam, Global's first major inland marine construction project.

Right: Tim Beaver leads the huddle during the salvage of the Viking.

Global was growing, getting lengthier gigs and more of them. Returning to the new office on West Marginal, Thom Davis recognized few people, as the company had hired more help to provide the infrastructure to go after these new kinds of work. One of three remaining founders who had been there at Todd Pacific Shipyard, spending day after day patching the rotten wood on sinking drydocks, he decided it was time to move on. He and his wife, Debbie, spent the next five and a half months in the mountains at Lake Cavanaugh, Washington, building a home. He still works in the marine business to this day, and doesn't see himself stopping anytime soon. "I'm sure I'll continue to do it for a long time after I retire. It's all I know," he said.

Heading into the new millennium, Tim Beaver and John Graham were the remaining founders. The company was rapidly expanding, in terms of both people and the scope of work it was willing and able to do. Global was going places—but where exactly, and what values were driving it?

SQG Provides Structure

The conversation started with Mike Langen and Devon Grennan. Grennan, who had just returned from a hiatus of his own, managing a yearlong oil spill response office in the U.S. Virgin Islands, said, "We sat down and thought that we needed to start talking about a business plan for the company, to get us all on the same page so we know why we're doing this and what it's for." Langen and Grennan brought their ideas to Beaver, Graham, and other key people, and around their busy schedules laid the groundwork for the SQG Program, standing for the company's core values: Safety, Quality of Service, and Growth.

"There were some long nights and drinks involved," Grennan

In 2002, a 63-year-old, recently restored Boeing 307 Stratoliner crashed into the shallows of Elliott Bay along Harbor Avenue, a mile away from Global Diving headquarters. While the operators were attempting to overcome a problem with the landing gear, the plane ran out of fuel. Global crews assisted in the salvage, and commemorated the occasion as the plane was being raised (at bottom). From left, Travis Havens, Tim Beaver, Ron Larsen, Dan Pierson, Dan Gilchrist, Rich Silvernale, and Jim Sherman (of General Construction).

Family

It's not easy having a diver in the family. Diving is a 24 hour a day, seven day a week profession that does not stop for holidays or special occasions. "You could be sitting down to Thanksgiving dinner, and if you get the call, you go," Jackie Lewis said. "It's rough on relationships."

Diving is a demanding career for both divers and their families. Husbands, wives, children, mothers, fathers, and other loved ones can expect the diver in the family to be gone on jobs around the world for lengthy periods of time, packing their bags at a moment's notice. They miss birthdays, school plays, anniversaries, and holidays.

However, diving is a passion, and many divers are fortunate enough to have support from their families to pursue their careers. "My wife was very supportive of me taking my career to whatever level I wanted to," Dan Pierson said. "She didn't always like it, but she understood when I got a phone call at one in the morning and might not see her for a month. It's great to have the support of a spouse like that, that lets you spread your wings."

At left, Devon Grennan's son, Jack, handles topside duties. Right: A drawing of Dad at work by Tim Beaver's daughter Margo.

recalled. "But what SQG really did was devolve some of the structure directly from Tim and John. It got a larger group involved to have a forum and talk about the direction of the company."

"We really felt we could shape the company," Langen said. "We felt like if we didn't shape it, nobody would." Langen credits Beaver and Graham's managerial style for fostering an environment where he and Grennan could head a program guiding Global's direction and defining its core values. "They're always there to help if you need it, but at the same time, they give you a lot of latitude and have a lot of faith in you. Tim and John took some considerable risks during Global's history and made some great decisions along the way that were the catalyst for our growth, but to be able to sustain that and not crash and burn was what motivated laying the foundation for the SQG program. Now everything we do comes from that."

SQG was rolled out at the company Christmas party at Salty's restaurant in 2001. Its effect is still felt 12 years later. "We did grow, we did give more opportunities for people, and there's more opportunity to bring people into the fold on various projects," Grennan said. "We know if we do those things [safety, quality, and growth] well, that will make us a better company."

Global had ample opportunity in the immediate future to test their values on two high-profile saturation dive projects.

Safety: Making 3-D Models to Mitigate Danger

In 2002, Global divers headed south to San Francisco for the company's second sat job. The SS *Jacob Luckenbach* sank 17 miles west of the Golden Gate Bridge after a collision with another vessel in 1953. She took Jeep parts, railroad equipment, and more than 450,000 gallons of oil to the bottom with her. Mystery oil spills washed ashore throughout the 1990s after winter storms and other episodic weather events, threatening wildlife in the Gulf of the Farallones National Marine Sanctuary, one of the world's largest marine sanctuaries. An investigation by the U.S. Coast Guard ensued, with the *Luckenbach* wreck a prime suspect. Using samples taken by two volunteer divers, the "fingerprint" from the *Luckenbach*'s oil was matched to that oiling birds along a 200-mile stretch of shoreline, and a multimillion-dollar oil-removal project began in 2002. The old wreck was going to be a challenge, located in the unprotected Pacific Ocean, miles offshore from the Golden Gate Bridge amidst vessel traffic lanes, and in shark-infested waters to boot. Safety was paramount in the minds of the dive team.

As the only West Coast dive company with sat and salvage capabilities, Global Diving partnered for the first time with Titan Salvage on the project. Crowley Maritime and PCCI, Inc., were also partners. Global has since worked with Titan on numerous projects. "That was our first exposure to a large salvage company, and they were a great bunch of guys to work with, very motivated, very job-focused," said Dan Pierson, one of the divers in sat for the job.

It was also Global's first effort at 3-D modeling and rendering, which both helped improve safety and added value for Global's clients, which included the Coast Guard and the National Oceanic and Atmospheric Administration (NOAA). Joanna Hawkins, a hydrographer, was hired to handle the positioning and diver location tracking. Using a computer model of the vessel provided by PCCI that was based on sister ship drawings, as well as multibeam sonar scans and data collected by remote operated vehicles (known as ROVs), her

The SS Jacob Luckenbach *was on her way to deliver supplies for the Korean War effort when she was struck by the freighter* Hawaiian Pilot *while in thick fog. The* Luckenbach *sank within 30 minutes.*

unique method of data interpretation was able to tell the dive team that the vessel was actually broken into three pieces, how far down in the mud it was buried, and at what angle it was resting. Usually with multibeam images, there is room for quite a bit of interpretation; by overlaying the multibeam scan on the computer model, the amount of interpretation was minimized. They now had an actual scale model that they were able to move and rotate in all directions. The divers now knew what they were up against before diving 200-some feet below surface, and Global could also give its clients a deliverable so they could see, understand, and visualize what was happening underwater.

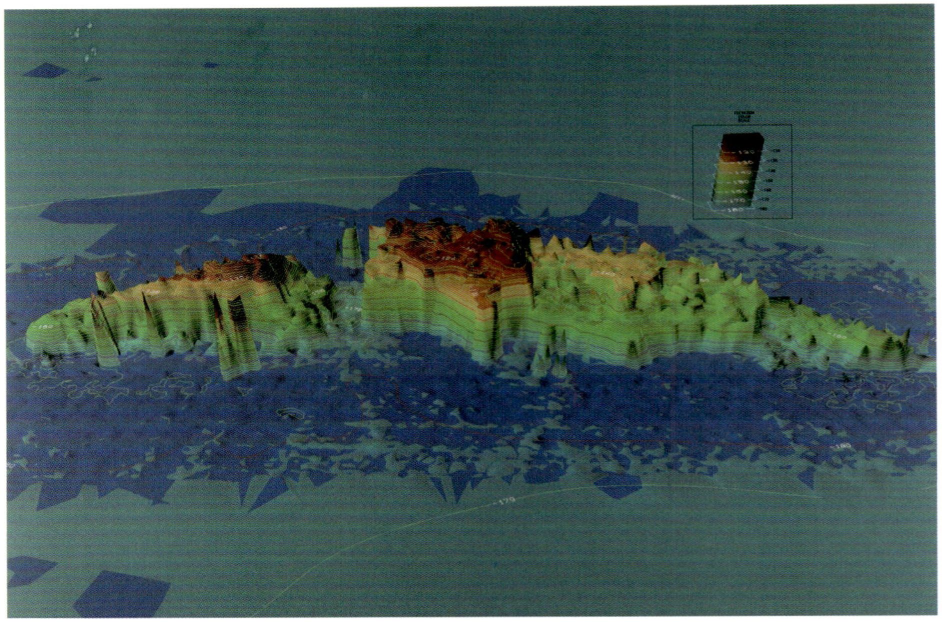

Above: A bathymetric survey 3-D model of the Luckenbach *wreck.*

Opposite: Global flies the flag proudly at sunup over the Luckenbach*, resting 200 feet below.*

To further improve safety, divers worked only 12-hour shifts. The *Luckenbach* rested in a busy shipping channel, and so the Coast Guard allowed work only during daytime hours. "We thought, You've got to be kidding me! This is great!" Pierson said. "Usually with sat it's 24/7, push, push, push, because sat is not cheap." The start of the project was delayed due to inclement weather. When the job finally was able to start, the oil-removal team had to figure out how to pump out thick, viscous oil the consistency of peanut butter. PCCI, Titan, and Global developed a method where steam was run through copper pipes to heat the oil tanks. A system of hot taps (allowing the divers to drill into the oil tanks while keeping the oil contained), valves, and pipes was constructed that recirculated the heated oil back into the tanks, making the oil pumpable prior to pumping it to the surface. By the end of the 110-day project, over 80,000 gallons of viscous black bunker oil had been removed.

"There's times when we think, If we don't do this, we're going to be left in the dust," Devon Grennan said. "The *Luckenbach* was one of those moments."

GLOBAL DIVING

As fate would have it, the *Luckenbach* provided the basis for two key relationships for Global going forward: with Titan, who would time and again bring Global out on salvage jobs, and with Capt. Scott Powell, an expert in offshore oilfield construction, marine firefighting, oil well blowouts, and salvage, who wrote a technical proposal for the project that, though unsuccessful, included Global in the plan. In typical Global fashion, the meetings were happy coincidences that several years later would change the company forever.

Personnel ready the 15,000-pound valve with actuator at Gross Dam.

Quality of Service: Staying Put to Get the Job Done at Gross Dam

The *Luckenbach* oil removal was Global's first sat project working under regulatory agencies, as opposed to the private investors behind the *Brother Jonathan*. NOAA and the Coast Guard had strict standards in terms of safety and the outcome of the project, and Global passed with high marks, giving the company the resumé needed to get an even larger sat job. Building off the success of the *Luckenbach* and the inland construction work completed at Rocky Reach, Global became the prime contractor to install a hydraulically actuated gate at Gross Dam, in Colorado, for the Denver Water Board. Denver Water had a lease for the dam with the Federal Energy Regulatory Commission (FERC), which mandated that upon its signing, Gross Dam would have hydro generation capabilities within 50 years; time was up, and FERC could choose not to renew the lease if the dam did not meet that stipulation. FERC leases were routine throughout the Midwest, and as more similar prerequisites came due, Global found more work on inland dams.

Denver Water's original idea was to drain the dam for construction, but a series of droughts forced them to scrap that plan. They needed divers to work at 285 feet below surface, and Global was the only outfit outside the Gulf Coast equipped to do so. Divers worked 24 hours a day for 55 days installing and grouting concrete jacking pipe in an inlet tunnel. The installation had to be exact: "We only had an inch to work with on each side," Tim Beaver said. Mike

Langen devised a plan where the divers laid track in the tunnel for a hydraulically driven cart that would transport the concrete jacking pipes into the inlet tunnel one at a time until they protruded from the tunnel face. Then, the divers installed a 15,000-pound stainless steel valve with hydraulic actuators.

The hydraulic controls for the valve were connected to the surface by stainless steel piping that ran down the face of the dam and across the lake bottom to the valve. Langen's plan called for surface gas divers to work in conjunction with a sat team. "We took various aspects of our expertise and brought these together in a way nobody else could," Beaver said. Though it was an ingenious design, Langen underestimated the difficulty of laying pipe across the steep and irregular terrain between the dam and the inlet works. The Colorado winter also proved difficult to deal with and what was supposed to be 25 days stretched out to 45. The surface divers worked much slower installing their pipe than the team in sat, and so the sat divers got restless. On the verge of a mini-mutiny, Langen persuaded them to persevere and finish the job. They were brought up after 55 days in sat, a Global record for one sat team.

Global's barge stationed at Gross Dam, high in the Rocky Mountains.

In an effort to deliver a better product to their clients at Denver Water, Global placed an emphasis on communicating the progress of the pipe laying, drilling, and grouting, through written updates and a bevy of accompanying images. "We're a company of pictures," Devon Grennan said. "If you go to our office, there's a bunch of pictures on the wall. The same goes for deliverables for our clients. We're trying to give real-time, accurate accounts of what's going on."

"Diving means the client can't be there," Tim Beaver added. "You have to be able to assure them through video, good reporting, and testing that the job is being done properly."

Growth: "Putting Meat on the Bones"

Of course, before Global could get the job, they had to ensure that they could complete the project in the first place. The Gross Dam project required a surety bond, which guarantees to the facility operator or owner that "if this job goes south, it will still get done," Beaver explained. "They find someone else to finish the job, then they take your house, car, business, tools, everything you have to pay the other company to finish. You don't ever want something to go to bond." As Global began to tackle multimillion-dollar projects for government entities, bonds became more commonplace. However, securing a bond requires a lot of paperwork, a lot of meetings, and a lot of time that divers managing projects do not have. Global Diving needed more behind-the-scenes help in the office.

One of the first major hires was Trinity Ng-Yeung, who joined in 2004 as controller and soon became vice president, heading up accounting and human resources, and leading the administrative support staff. When she started at Global, the accounting team was Jackie Lewis. "Jackie ran everything," Ng-Yeung said. "She was the accountant, the bookkeeper, and the Mom. They needed someone with accounting training to take on the growing load. This seemed like a company that had the proper skeleton and needed new skillsets to put the meat and muscles on, to make it ready for the next stage."

Ng-Yeung's background in both the public and private accounting sectors taught her that the bottom line drives operations. That's not always the case at Global, as the company's founders made clear early on to their new employee. "When Tim and John told me that, as an accountant, I thought, do you really need me here then? Slowly, I understood what they meant. We as a company think first of bringing our employees home safe, even if it costs us more."

Ng-Yeung became a vital addition to Global, helping put "the meat on the bones" and developing a support staff over time that increased Global's project capacity. Now, Global was about to venture into uncharted territory for the company, heading to the Gulf Coast to embark on a project that would be their greatest test of the commitment to safety, challenge their abilities to deliver a quality product, and allow the company to grow in ways previously unimaginable.

Crews launch the bell for the first run on what will be a Global record 55 days in sat for one team.

Seeing from the Surface

SINCE GLOBAL DIVING'S INCEPTION IN 1979, "a lot of things haven't changed a bit," Tim Beaver said. "We still use the same exact hoses, the same exact compressors, the same methods. Suits and hats have changed relatively little.

"But one thing that's really improved is the dive cameras, and they've made a huge difference across the board for divers."

Today, lightweight, compact, and relatively low-cost dive cameras that send instant, high-quality video streams to the surface are mostly taken for granted as a standard, trustworthy technology.

Not so when Global began work. The first dive cameras were "Frankenstein lash-ups" made from home video cameras and encased in homemade housing. The cameras had an on-off switch, a focus switch, and a light switch that was placed in a plywood box on the surface. The camera itself was protected inside a welded-shut aluminum pipe. The housings were prone to leaks, the cameras to shorts, and the focus to not working, when the rubber band attached to the surface-control motor slipped off or broke. Many of Global's contraptions came from "Horrible Hal" Iverson's workshop in Seattle's University District.

As camera technology has improved, the demand for divers has increased, and Beaver believes this is no coincidence. "An engineer must have real confidence that his project is being built well and to specification," he said. "When this confidence is there, the work follows. Now divers are used more than ever on critical, highly technical projects, and this is in no small part due to smaller, better, and more reliable cameras."

DIVER
RESCUE
GLOBAL

chapter 9

A Salvage Mentality: Global Enters the Gulf of Mexico

Capt. Scott Powell knew that divers were a strange breed, but the crew piling out of the rented minivan appeared to be true outliers. Capt. Powell was vice president of Houston-based Wild Well Control's Marine Division and worked with offshore oil clients around the world, literally putting out fires, as well as assisting with other maritime emergencies. Now he was in the Gulf of Mexico, dealing with the aftermath of Hurricane Ivan in 2004, in which his client, Noble Energy, had lost three major oil platforms with 77 live wells shut in (live and flowing but held in check by an emergency safety valve). To remove the downed platforms and ultimately control and kill the wells, he needed saturation divers with a strong salvage background, something he could not find in the Gulf, and so reached out to the Pacific Northwest for Global Diving's help.

Before lengthy jobs at sea in the Gulf of Mexico, Capt. Powell always made a trip to the Galveston, Texas, Walmart to stock up on DVDs to keep his workers entertained. On his drive back to the *Boa Deep C*, a brand-new 400-foot DP3 vessel that would be going out on this crucial and dangerous job, he spotted a minivan. It was nearly riding on its shocks because it was so loaded with passengers who were large, grown men wearing souvenir animal and fish hats, looking like they'd just come from Disneyland. One man in a shark hat hung his

"See any fish down there?"

THE MOST COMMON QUESTION ASKED OF A DIVER

Global's Sat IV on the deck of the Superior Performance.

arm out of the open side door as they barreled down the freeway. That man was his lead diver, Don "Hoss" Hosford, and Capt. Powell immediately knew that this was his diving team from Global. Though they may not have looked it, they were ready to get down to business on what would become a momentous turning point in Global Diving's history.

Below: The downed ENSCO 64 platform following Hurricane Ivan.

Bottom: Hurricane Ivan making landfall on September 16, 2004.

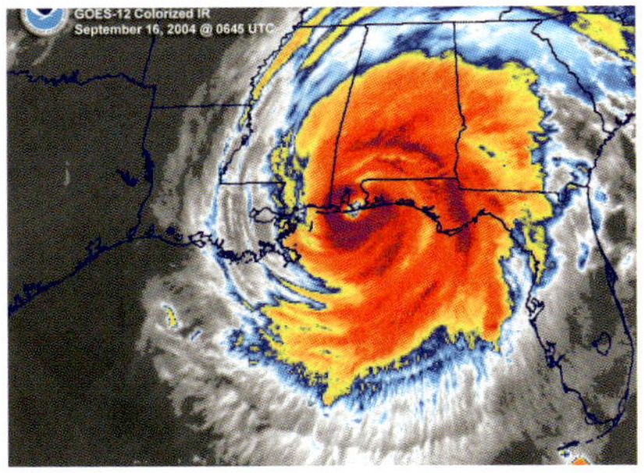

Gulf Recoveries

Though Capt. Scott Powell first became acquainted with Global Diving on the *Luckenbach* oil recovery project, it was on another Ivan-related salvage that the divers from the Northwest really caught his eye. ENSCO, a London-based gas and oil company, had hired Titan Salvage to recover three expensive high-tensile steel legs resting 300 feet underwater 83 miles southeast of Venice, Louisiana. The legs had held up the ENSCO 64 drilling rig until the platform found itself directly in Ivan's path and was battered by 83-foot waves. The uprooted legs and the downed platform drifted some 40 miles away from their drilling location.

Titan hired Global to assist in the recovery. An upgraded Sat I, with rebuilt launch and recovery system, was readied for the first time on a large vessel for some heavy-duty offshore work. The job was never completed—not all the legs were recovered due to client priorities—but Global made an indelible mark on Capt. Scott Powell, who oversaw work on one of the legs that was recovered next to a live oil wellhead. "I was impressed by the way they were conducting their business," he said. "They had direct ownership involvement from Tim and John in their day-to-day projects, and that really caught my attention. That was something that was extremely key to me."

When Noble Energy called with news of the damage caused by Hurricane Ivan—Capt. Powell was overseas seeing to a different well fire and blowout—he knew he wanted Global on the project, regardless of their lack of experience working on downed platforms. "We were fixing to venture into some of the harshest salvage conditions in the world. These downed platforms were loaded with

Massive chains were used to raise the uprooted ENSCO 64 leg resting 300 feet below surface.

To improve safety for Global divers and allow work to continue 24/7, Wild Well Control enlisted the help of Hollywood lighting guru Pete Romano to create better lighting for the work site. The end result beamed 45,000-48,000 candlepowers of blue-black light into the pitch-black waters. Though skeptical at first, the divers soon declared they'd never dive without it again. "I consider the development of the lighting system to be a major success between us and Global," Capt. Scott Powell said.

stored energy that we knew we had to arrest to keep the divers safe. We had to have the heart of the owners directly involved to make this a successful, safe program." Global's unique capacity to perform saturation diving with what Capt. Powell called "a salvage mentality" also made the company a good fit for the unprecedented project that was shaping up to be the largest salvage job ever done in the world up to that time.

Capt. Powell called Tim Beaver, leaving a message that he had a major project to discuss with him. He didn't receive a call back. After two more attempts, he finally got hold of Beaver, and urged the Global president to take a leap of faith and fly down south. Beaver brought with him a 15-slide PowerPoint presentation and one suit and tie, just in case, though he figured he would be meeting at "a series of double-wide trailers out in a field somewhere with bulldozers, fire monitors, and big fancy pickup trucks in the parking lot." Instead, Beaver was buzzed into the former Chevron Oil Drilling Research Center, where he was greeted in the lobby by giant plasma-screen televisions looping footage of fires, blowouts, and other mayhem. The facility had a fully functioning oil well for training purposes. "I was glad I wore that coat and tie," Beaver said.

A Global diver walks on conductors after a well survey.

In the meeting with Capt. Powell and members of Noble Energy's board of directors, safety became the primary topic of conversation. Based purely on the expected number of wet man-hours for the project, they estimated that there would be 2.4 dive fatalities on the job. "They said, 'That's not happening,'" Beaver recalled. One of Noble Energy's board members had been involved in a diving fatality, and was adamant that there would be none. "That suited us right to our toes; there's nothing better than having a client that's into a safe project first and foremost, and willing to take the time and expense to plan around the risks ahead of time."

Global Diving signed on to join a crew of other subcontractors. Capt. Scott Powell stuck his neck out on the decision to hire a dive outfit from outside the Gulf. "Some Gulf dive companies were really nipping at my heels: why did I bring a company from Seattle all the way down to the Gulf of Mexico? My defense to that was, I had a group in Global that already had success in salvage work. They knew when the hair was going to stand up on the back of their neck not to do something, and that's what's most important in salvage—to know where not to go. I didn't want oilfield guys; I wanted salvage divers who could do sat."

Breaking Doors Down

For Bruce Humberstone, Dan Pierson, and others involved, the new work in the Gulf was a triumphant return of sorts to where they got their start in diving. "We were going back to where we all started. Even better was going down there with people telling you you've got no business being there," Pierson said.

They also were burdened with the specter of 2.4 fatalities estimated over the duration of the job. "It was a little bit intimidating, knowing that stat about casualties ahead of time, but none of us ever looked at it that way. It was just another project: let's go prove these guys wrong. And that's what we did."

After stowing their shark hats for post-project celebrating, the dive team boarded the *Boa Deep C*. They brought with them a rebuilt Sat I, refurbished once more for improved safety on what was sure to be a lengthy job. Capt. Scott Powell called a meeting with all the various teams of contractors, many competitors, in total numbering 100 individuals. The first thing he required was that everyone retire their company logo T-shirts. "We are the hurricane recovery project, and that's where our hearts are going to be," he told them. "You can go back to your companies later. We're living on this boat, and we're going to be a family."

Top: Left to right, Pat Boyd, Joe Carp, Dave Partlow, Jim Geissler, Nick Finney, and Brian Jamison.

Bottom: Dave Partlow, left, and Bruce Humberstone led the charge in the Gulf during the recovery efforts.

Top: The Boa Deep C.

Bottom: Don Hosford plays tic-tac-toe against an ROV. Photos and video of the unprecedented contest reverberated throughout the diving industry.

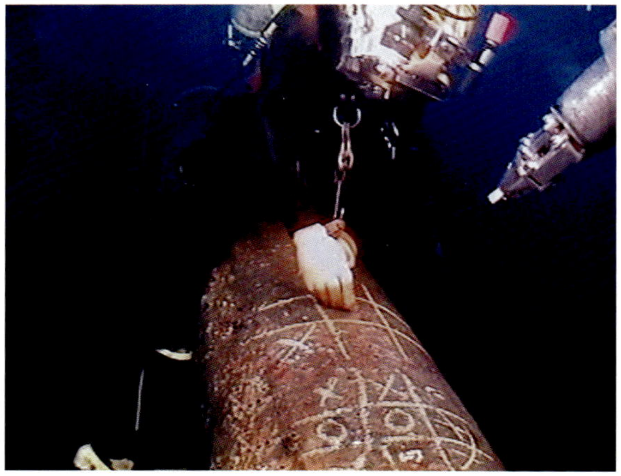

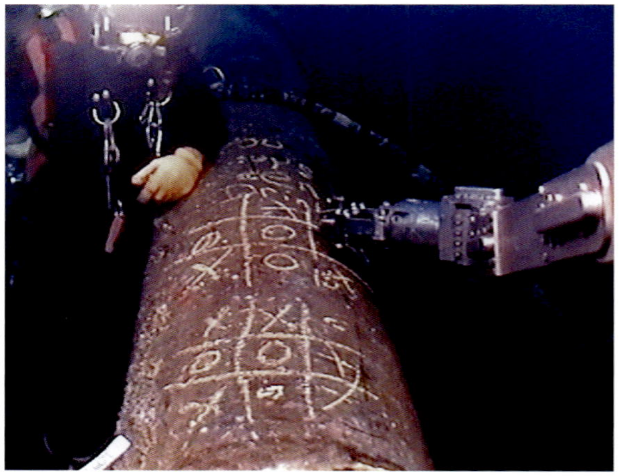

As in any loving family, there were some hijinks—in particular an ever-escalating sticker war between Global and ROV operator Oceaneering—and some unexpected cooperation. Eventually a diver and ROV worked jointly underwater, something that had seldom been done before and was considered risky. A video of a diver playing tic-tac-toe against an ROV immortalized the achievement and reverberated throughout the dive industry.

Building Sat II

The hurricane recovery team finished one out of three platforms before Hurricanes Katrina and Rita hit barely a month apart from each other in 2005, causing further devastation. The calls kept coming in—there were now 47 downed platforms.

For the Global team to do sat on a project of this scope with such a strict emphasis on safety required a major upgrade in their saturation diving capabilities—an entirely new system, in fact. Bruce Humberstone and Tim Beaver made the 11-hour flight from Seattle to a warehouse in Norway to inspect an Italian-made Drass system from the 1970s, which they bought on the spot for $900,000. The system was shipped to the company's newly purchased fabrication workshop in Seattle and a multimillion-dollar renovation began.

The financials of building the system and funding the upcoming Gulf project were new territory for Global. "When you build sat systems, a few million each, you're in a different ballpark," Trinity Ng-Yeung said. "We needed more in-depth information for the banks. They want to see your work in progress, if you have a healthy balance sheet and the operational integrity and savvy to help mitigate their risk—you have to prove to them it's okay to lend you the money."

Global Diving, beginning with Dworshak Dam in 1991, built capital in an atypical manner. "Capitalized companies say, Let's build gear and go find work for it," Beaver said. "We could rarely afford to work on that model. We had to know what we could do, be ready, and when the opportunity came, make a decision fast."

Top: Sat II being built in Global's Seattle warehouse.

Right: Completed Sat II on board the Ocean Carrier.

The decision to build the system that would become Sat II did arrive quickly, and left Humberstone with little time to make the necessary modifications. He parked his trailer outside the shop and lived out of it for two months while working 14-hour days, only making the hour-long commute to his home in Carnation, Washington, on the weekends. "To take a bunch of junk you buy over in Norway, ship it back here, figure out how to make it all work, and add what you have to add to it is quite an operation," he said. "It was especially quite an undertaking for a small diving company."

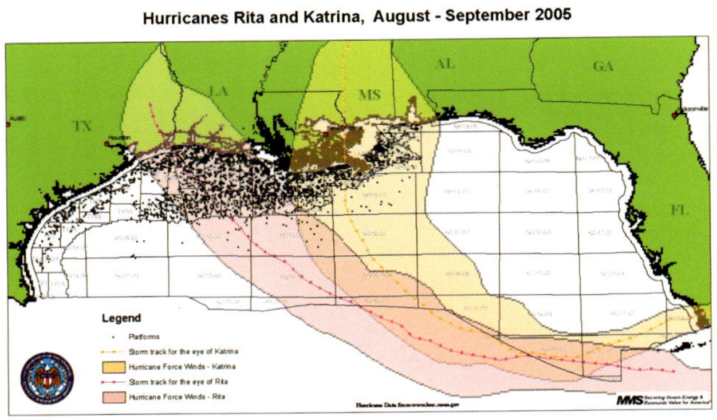

After Hurricanes Rita (pink) and Katrina (orange) in 2005, Global's workload in the Gulf increased exponentially. On this map, each black dot represents a platform.

"In It for the Long Haul"

"During the previous year, we had developed new technology, business models, policies, and procedures," said Capt. Scott Powell. "With that in place, Global was instrumental in laying the groundwork that cleaned up the Gulf of Mexico." With the workload exponentially increased, Global Diving was ready to stay in the Gulf for the long haul. "They were right there to the bitter end. I've seen a lot of these divers literally grow up, guys that didn't have gray hairs then that do today."

Beaver and Graham were present for much of the operation, directly overseeing 980 bell runs, 3,300 dives, and over 140,000 dive hours—and without a single time-loss incident. The divers continuously impressed, with both their talents and speed. "Nobody could believe how West Coast divers could burn and rig as fast as we did," Pierson said. The divers set production records, even as engineers requested that the divers slow down so they could keep up with modeling and completion drawings.

Global was now firmly established in the Gulf of Mexico, and had more business than the founders could possibly have imagined 25 years earlier. "Finally, after all these years, we had the capital to build the kind of company we thought we could build," Tim Beaver said. And now, with their resumé, infrastructure, and equipment, they could tackle a project that would take Global divers deeper than ever before.

Global Milestones

West Marginal Way office, opened 1997

The Seattle warehouse

Mare Island, California, office, opened 2012

The warehouse at Mare Island

1979 Founders join and open office on Harbor Island

1980 Global incorporated

1984 First brand-new truck

1985 Tug *Spike* purchased

1988 First full-time employee guaranteed 40 hours a week of shop time with benefits

1990 Tug *Spar* added to fleet

1990 Oil Pollution Act passed, which revamped the spill response industry by creating hard liability standards, a multibillion-dollar cleanup fund, and strict pollution response standards

1992 DSV *Prudhoe Bay* acquired

1997 First saturation system, Sat I, commissioned

1997 Move to West Marginal Way SW

1999 First major marine construction project

2002 SQG (Safety, Quality of Service, and Growth Initiative) to focus and manage growth of the company, while preserving the quality of service and high safety standards the company employed since its inception

2004 Inshore Divers, Inc., in Rio Vista, CA, acquired by Global

2004 Hurricane Ivan brings Global to the Gulf of Mexico

▲ **1979**
Raymond J. Bushey
Seattle, Washington

1980
Todd Pacific Shipyard drydock work
Seattle, Washington

▲ **1981**
Todd Pacific and Lockheed shipyards naval frigate shipbuilding contracts
Seattle, Washington

1984
The *Warrior* salvage
Blaine, Washington

1984
Texaco oil refinery project at
Anacortes, Washington

1985-86
***ARCO Anchorage* oil spill response**
Ediz Hook, Port Angeles, Washington

1988
***Nestucca* oil spill response**
Grays Harbor, Washington

1989-90
***Exxon Valdez* oil spill response**
Valdez, Alaska

▼ **1991**
Dworshak Dam
Clearwater County, Idaho

1991-92
***Tenyo Maru* oil spill response**
Neah Bay, Washington

1994
Salvage of the *Jupiter*
Seattle, Washing

200 ft
400 ft
600 ft
800 ft

80 90

▲ **1999**
Rocky Reach Dam construction project
Chelan County, Washington

▼ **1999-2000**
SR-520 bridge cable replacement
Seattle/Bellevue, Washington

2004
Canyon Ferry Dam spillway basin repair
Helena, Montana

▼ **2004**
Gross Dam
Boulder County, Colorado
285 ft

2003
Corner Collector surface bypass installation
Bonneville, Washington

1996-97
Wyckoff remediation
Eagle Harbor, Washington

▲ **1997**
***Brother Jonathan,* round one**
Crescent City, California
285 ft

1998
Alyeska Pipeline anode system installation
Valdez, Alaska

▲ **1999**
***New Carissa* salvage and oil spill response**
Coos Bay, Oregon

2000
***Brother Jonathan,* round two**
Crescent City, California
285 ft

◀ **2002**
SS *Jacob Luckenbach* oil removal
Gulf of the Farallones, California
200 ft

Harbor Island office, opened 1979

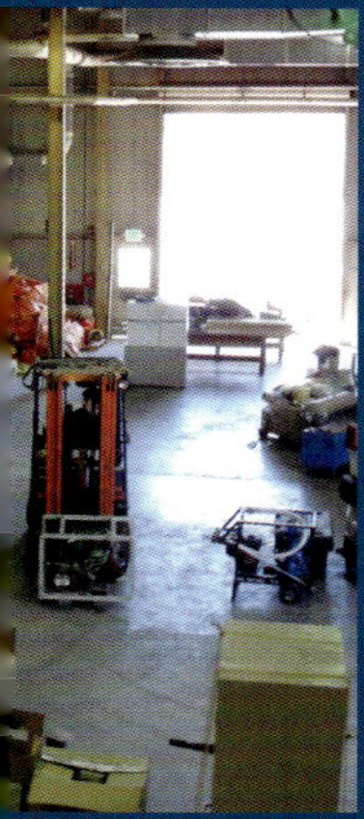

The Anchorage, Alaska, office, added to the Global family in 2009

Global motorcycle created by Chris Drakos, longtime employee and owner of Odd Cycles

The Houston, Texas, office, opened in 2010

2005 Hurricane Katrina expands Global's role in the Gulf of Mexico

2006 Sat II acquired, modified, and retrofitted

2007 Sat III commissioned

2007 Anacortes, WA, auxiliary office opens and preventative boom begins for Tesoro facility

2009 SAAB Cougar purchased for Robson Bight

2009 Global acquires Offshore Divers of Anchorage, extends footprint in Alaska

2010 Sat IV commissioned

2010 Houston office opens

2010 Richmond, CA, office opens – expanded Enviro Operations in the Bay Area

2011 Idle Iron policy increases demand for Global's services in the Gulf of Mexico

2011 Tim Beaver and John Graham receive the Ernst & Young Entrepreneur of the Year® 2011 Pacific Northwest Award in the Transportation and Logistics category

2012 Rio Vista and Richmond offices combine on Mare Island, Vallejo, CA

2012 Strategic emphasis on Global's core service lines: Marine Construction, Engineering & Technology; Casualty & Emergency Response; and Offshore Operations

2010
Deepwater Horizon oil spill response
Gulf of Mexico

2010
***Princess Kathleen* oil removal**
Juneau, Alaska

2010
Pride Wyoming rig salvage
Gulf of Mexico

2010
***Princess of the Stars* hazardous material recovery**
San Fernando, Romblon, Philippines

2010
Cheesman Dam
Douglas County, Colorado
200 ft

2010
Alyeska Pipeline mooring buoy refurbishment
Valdez, Alaska

2011
Crescent City tsunami response
Crescent City, California

2011
***SS Montebello* ROV oil assessment**
Monterey Bay, California
900 ft

2012
***Christy Lee* fender repair**
Cook Inlet, Alaska

2012
Tug *Tiger* response and salvage
Richmond, California

2012
***Costa Concordia* salvage**
Isola del Giglio, Tuscany, Italy

2013
Horse Mesa Dam
Phoenix, Arizona
260 ft

2013
Pacific Gas & Electric North Area Dive Bundle
California

200 ft
400 ft
600 ft
800 ft

▲ **2005**
ENSCO 64 leg retrieval
Gulf of Mexico
300 ft

2005
Noble Energy
Gulf of Mexico

▼ **2006-2007**
SS *Catala* oil removal and salvage
Ocean Shores, Washington

▼ **2008**
The Dalles spillwall repair
The Dalles, Oregon

▲ **2008**
BP Properties Platform recovery, GI 47 C
Gulf of Mexico
120 ft

▼ **2007-2012**
New York City water supply project
New York, Shaft 6
720 ft

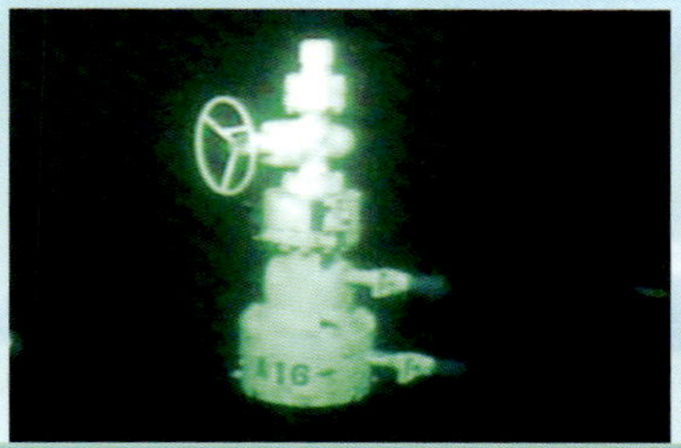

▲ **2009**
Wild Well Control well head installation
Saudi Arabia

2008-2009
Robson Bight salvage
Vancouver Island, Canada
1,165 ft

TEAM

The Global team prepares to launch the bell.

chapter 10

700 Feet Below New York

The diver cranked on the wrench he affixed to a bolt holding in place a piece of 24-inch bronze piping. He would have to loosen several more bolts before the piping came free and he could begin work removing the next section. This was a routine task, the likes of which he had done countless times before and would do many more times on this particular job, but whether he acknowledged it or not, this job in 2007 was different. There were more eyes on every turn of the wrench than ever before, and a whole lot more was at stake—for his company, Global Diving; for their client, the New York City Department of Environmental Protection (DEP); and for 15 million people, the vast majority of whom would in all likelihood never know of the work taking place 700 feet underground in Shaft 6 of the Delaware Aqueduct system.

This job, dubbed Shaft 6 by the Global team, was the one that would require all the expertise, equipment, and personnel the company had attracted and accumulated—much of it by happenstance and good luck—over the past 27 years.

All Hands on Deck for Bidding, Bonding, and Estimating

The New York City water supply system is a modern marvel and a source of pride for New Yorkers, who boast that they have the best drinking water in the country. Built in the early 20th century, a series of aqueducts (some as deep as 1,200 feet underground) transport water from watersheds as far as 100 miles away. Forty percent of the water flows from reservoirs in the Catskill Mountains along the

Left: Gas cylinders are lined up and ready to go for Global's deepest job ever.

"Man, that water looks cold."

A COMMON, WRY COMMENT FROM TENDER TO DIVER MOMENTS BEFORE A DIVE

45-mile-long Rondout-West Branch tunnel straight to faucets in the five boroughs, as well as to nearby suburbs. Daily water consumption averages 1.2 billion gallons.

In the late 1980s, the Rondout-West Branch began leaking, to put it mildly. By the early 2000s, 20 million gallons a day gushed from the aqueduct, flooding basements on otherwise perfectly dry days in the neighboring hamlet of Wawarsing. On its worst days, some 36 million gallons leaked. Sinkholes appeared in backyards. Maintenance plans called for draining the aqueduct to do work in the dry, but with so many millions dependent on the system for water, that simply was not an option, given the extended periods of time needed to investigate the condition of the tunnels and repair accordingly. This fix would, at least in part, require diving.

Global veterans fondly recalled "all-hands" projects in the early days, those important, large-scale jobs that required the whole team to be out of the office and in the field; this was an all-hands job well before Global had even won the bid, though one in which the support staff could stay put in Seattle. Global Diving & Salvage set about producing the largest single bid the company has ever put out (as of this writing). Global was looking to be the lead subcontractor for the project, but had to demonstrate to the potential prime contractors that, 1. they were capable of taking on a job of this scale and had the experience to prove it; and, 2. they had the bonding and financials necessary to see the project through.

First, Global submitted a technical proposal to be vetted for the job and put on a prequalified list of finalists. They got on the shortlist and were the only dive company not from the Gulf of Mexico. "That was a big deal for us. Here we are, this little dive company out of the Pacific Northwest, and suddenly we're on the map," said Mike Langen, who would make numerous cross-country trips during the bidding of the project. He and David DeVilbiss, who had fortuitously moved to Ithaca, New York, shortly before the project arose, met with the prime contractor for the project, Rondout Constructors. Langen and DeVilbiss had with them a thick binder full of evidence showing that Global had the saturation diving and construction experience to do the job, with

The New York City water supply spans three aqueducts and 6,500 miles. The Delaware Aqueduct is the world's longest tunnel, at 85 miles long.

Opposite: Construction crews work in the dry aqueduct during its original construction in 1937.

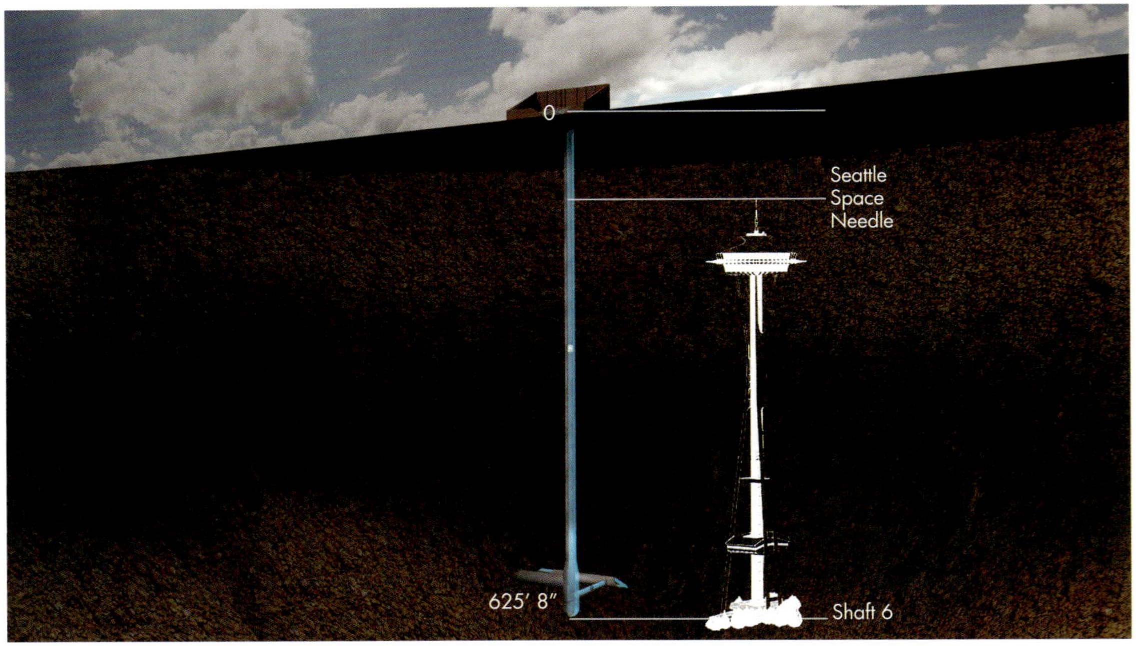

Global created 3-D models of Shaft 6. The computer-generated models were essential during a project in which the vast majority of the people and agencies involved would never see, in person, the work done.

Gross Dam as their resumé centerpiece. Global's bid for $20 million, leaps and bounds above the value of the largest projects the company had completed, was accepted. "It was a staggering number for the company," Mike Langen said. "Even before we started work, some real milestones for the company were set."

One important milestone, and one factor contributing to the high bid, was that Global would need to build a new sat system. Sat I and Sat II were being used on other projects and would likely be unavailable for Shaft 6, so a new system—which would become Sat III—would have to be built specifically for the project. Once again, Tim Beaver and Bruce Humberstone searched worldwide for the parts and pieces necessary, and found them during a whirlwind six-day excursion that took them to London; Norway; Copenhagen; Edinburgh; Washington, D.C.; and Aberdeen, Washington. The lot was immediately shipped to New York and transported by rail to Seattle, where modifications were already being engineered in the Global shop.

Back in Seattle, another team mobilized to secure bonding. "Shaft 6 is one project that really impacted the entire company, from accounting to operations, because it required a large bonding capacity," Trinity Ng-Yeung said. "For us to get that large a bond, it's not like we just say, 'Hey, insurance, give us a large bond.'" Ng-Yeung remembers Tim Beaver, John Graham, Devon Grennan, and Mike Langen having to "put on suits" to visit the underwriters and educate them about the project and Global's capabilities to complete it, from both an operational and financial standpoint.

The meetings were a learning experience for the Global owners and managers. "Emphasizing the 720-foot depth of the dive seemed to make the bonding companies nervous," Beaver said. In later meetings, they never mentioned the depth unless asked, instead focusing on the aspects of the job that Global had lengthy experience in. "Then when we were asked about the depth, we learned to say 'just 700 feet,' like that depth would be a walk in the park."

Through in-person meetings and stacks of documentation, Global was able to convince the underwriters that the company had the expertise and experience matrix to take on projects of this size and

Entrance of Shaft 6 with Sat living chamber seen from outside.

Top: Personnel build the Shaft 6 mock-up.
Bottom: Dive helmets equipped with the new rebreathers.

scope. They proved that Global was ready to take the leap, and that the underwriters should take the leap with them.

When bonding was finally approved, Beaver and Graham flew back east for a last meeting at the AIG building in Lower Manhattan. Grabbing a quick bite before the meeting and wearing full suit and tie, Beaver spilled his vegetable beef soup down his fresh white shirt. Stripped to the waist in the bathroom, Beaver applied soda water to the stain, and the show went on. After a polite meeting with several bonding types, all was set to go for Shaft 6.

Preparations for a Big Step

As a condition upon acceptance of Global's bid, the company had to build a full-size mock-up of the 13-foot-diameter Shaft 6 and perform complete run-throughs of the tasks they would be performing at depth back east. When the real job started, the team would have to be on their game from the get-go—the DEP could only shut down the aqueduct for a limited window of time, so the run-throughs were essential for working out the kinks and making exact preparations for the work ahead.

Global performed the run-throughs in shallow water at Todd Pacific Shipyard, with representatives from DEP observing. Though this was an unusual bid requirement, the team found the experience valuable. "I wish every job had that requirement," Tim Beaver said. "I want to know before we get out there and have guys pressed at 700 feet whether the tools and procedures will work. It was great."

The mock-up job gave the divers a chance to adjust to working with three-man teams in the bell (diver, tender, and stand-by diver), contrary to the industry standard of two, an unexpected requirement from the federal Occupational Safety and Health Administration. Now a third body, with a third set of gear, had to fit into the bell. In addition, redundancy of all systems was a contract requirement, so a second bell had to be on standby and ready to be sent to depth to unfoul or recover the primary bell in the extremely unlikely event that it was required. Tim Beaver and Bruce Humberstone were able to find in

The complete mock-up is readied for a test run at Todd Shipyard in Seattle.

England two bells, one larger and one smaller, that could both mate to the system and were perfect for a rescue operation. Global also supplied a hyperbaric rescue chamber, or HRC, but launching the HRC at this unusual job site proved to be a challenge. Instead of a launch system, Global engineered a system of skids and forklift attachments so that in the event of fire, flood, earthquake, or other catastrophe, the divers could be removed from the site and decompressed safely.

The divers also learned how to use a new rebreather system for bailout at extreme depths. In case the air hose was cut or the diver lost gas, he could plug the rebreather into his helmet and have 20 minutes to return to the bell, as opposed to the four minutes he would get from a normal bailout, which is essentially a scuba tank. The new rebreathers cost $30,000 apiece and the company flew out a specialist from Scotland to train the divers on how to use them.

"Not NASA Enough"

After months of preparations, the time finally came for Global to make the initial bell run, amidst an unusual amount of fanfare for a dive job. The first phase of the project was to determine the condition of valves and pipes at the bottom of Shaft 6 so as to ensure that it would not suffer damage or even collapse upon its eventual draining. The DEP made a small party of the occasion, complete with popcorn and soda. As anyone who's seen dive work can attest, it's not much of a spectator sport. "It should've been more NASA, with more of a show, but what they got was a construction atmosphere," Tim Beaver said. He and John Graham were on hand for the first bell run. The on-site Global support team meticulously triple-checked all tools and equipment. When the 8-foot-diameter bell of Sat III was finally lowered down the 13-foot-diameter Shaft 6, the divers prepared to exit and begin work. Then the support team above water got a call.

"Topside, we have a little problem."

"What's that?"

"This bailout rebreather thing won't fit through the hatch when we wear it."

"Somehow, amongst about a million details that we had to cover in preparation, nobody ever thought about the diameter of the hatch and getting the new rebreathers in and out while on the divers' backs," Mike Langen said. The bell had to be brought back up. Upon conferring with the divers and DEP supervisors, it was decided to use twin 80-cubic-foot scuba tanks for bailout bottles in the interim, and to stage extra tanks strategically in case of emergency; for later phases, the hatch was enlarged and the rebreathers were used. The bell finally went back down late into the night, observed by a dwindling crowd of sleepy onlookers.

Over the first phase of the project the three-man teams rotated 12-hour shifts, with each diver working four hours at a time. In preparation for the eventual dewater and bypass of that section of the aqueduct for repair, the divers had to perform detailed inspection and measurements of the existing 5-foot-wide by 7-foot-tall horseshoe-shaped horizontal drift that connected the shaft with the main aqueduct. State-of-the-art sonar was used to obtain high-resolution measurements of the drift and shaft bottom area. Divers collected concrete core samples and performed ultrasonic testing on the three-inch-thick manganese bronze hemispherical submarine door that isolated the drift at the bottom of Shaft 6. The divers could work only in the dead of winter during the time of least water use by residents, so as to minimize the impact of shutting down the aqueduct.

Return Trips

Global teams returned to Shaft 6 for four separate sat runs over the next five years, completing the project in 2012. On the second visit, divers removed 25,000 pounds of piping, cutting it with a remotely operated rotary mill. Most of the equipment and supplies were specially designed by Global for the job. Construction Superintendent Dean Stone oversaw the fabrication of all custom tooling used during the project, and had a large role in planning and executing the work. Each task had to be meticulously planned so that anything to be used by the divers could be lowered down the 13-foot-diameter shaft

Top to bottom:

View of job site seen from above.

Removing the old gate valve.

Crew posing in front of new gate valve before installation.

Top left: John Lapeyrouse gets a view into the bell and divers A.W. McAfee and Spencer Dell.

Top right: Marc Mittry communicating with sat divers.

Bottom left: Left to right, Bruce Humberstone, Carl Damgaard, Mike Langen, and Dean Stone stand before a piece from the project.

Bottom right: A diver guides the blanking flange into place.

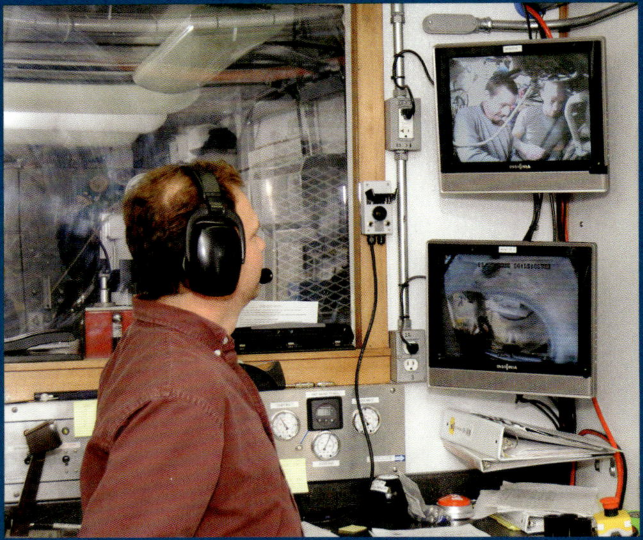

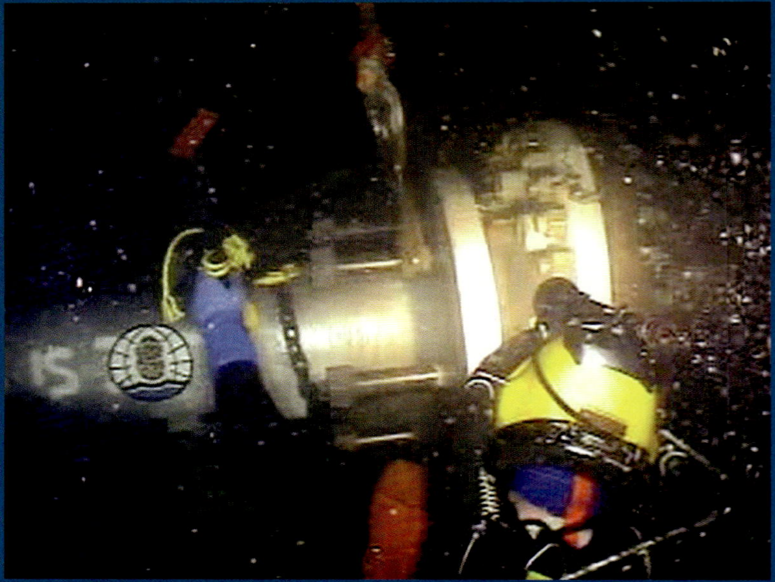

before the bell was deployed. Anything removed had to come up after the bell was retrieved. The team also removed the existing gate valve and installed a new 24-inch hydraulic stainless steel gate valve.

The most complicated task came when they installed a five-by-seven-foot, four-foot-thick mechanical plug weighing 23,000 pounds and designed to withstand 2.5 million pounds of force. The horseshoe-shaped plug would allow the shaft to be drained for more repairs. "The Claw," a remotely operated hydraulic lifting arm, designed and built by the Global team, assisted divers in lifting heavy components used in the assembly of the plug (some in excess of 500 pounds) so that they could bolt them in place.

Dealing with the many layers of bureaucracy on the project was initially a challenge, but once Global established their expertise and commitment to doing good work, the higher-ups on the project looked to them more and more for their input on how to get the job done right. "They were intelligent, professional men who understood the work, understood the various challenges that they may undertake, and they were willing to accept those risks," said Dan Peterson, Project Superintendent on the Shaft 6 project for Rondout Constructors. "They represented the work that they have done, and what they felt they could do, to assist our project in a positive manner." The highest compliment paid was when Global was asked to assist on repairs "in the dry," or work that did not require diving. The Global crew was so trusted that Rondout was willing to pay standard dive wages for the dry construction work (wages far and above those of construction) to ensure the job was done safely and right.

"Shaft 6 lifted Global to an elite status as a diving company," Mike Langen said. "There's nobody else out there that can say they've done that kind of job. It was a construction job at 700 feet, which was unheard of, period. After doing that job, we thought there was nothing we can't do, and I think our people believe that, and we certainly as a management team believe that."

And with a once-in-a-lifetime, company-defining project under their belts, Global Diving put that belief into action.

Top: Engineer Leif Stepakoff and Tim Beaver at the job site.

Above: Sat divers for plug install phase, left to right: Jimmy Geissler, A.W. McAfee, Dave Gill, Ryan Smith, Eric Bittick, and Chris Hackworthy.

VING
SCV-2
DIESEL

chapter 11

A Growing Family

David DeVilbiss arrived in Tacoma, Washington, to tackle his latest disaster. A cargo barge was on the verge of rolling over. DeVilbiss had hung up his drysuit for good and was now Global Diving's vice president of Marine Casualty and Emergency Response Services, managing projects for the next generation of Global divers. The company was prime contractor for uprighting the barge, and figuring out a way to do so before it completely flipped over was on DeVilbiss's shoulders.

Global got to the barge in time to stabilize it and bring it back to an even trim. It was during those stabilization efforts that DeVilbiss realized he had seen this same barge, in the same predicament they were trying to prevent, some 10 years earlier—but under different circumstances for Global. Back then, Global bid as a subcontractor under Fred Devine Diving & Salvage. The barge, then a self-unloading log carrier, had turned over completely while in Canada, so that the cranes that were on deck at the time were underwater and aimed at the ocean floor. DeVilbiss worked under the tutelage of Tim Beaver, who nurtured his interest in salvage work. The plan was to cut the cranes

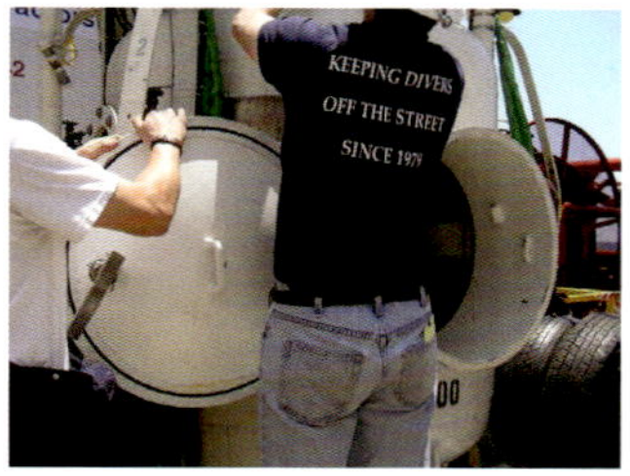

Left: *Preparing for a surface dive on the salvage of the* Albion *in the Monterey Bay National Marine Sanctuary off the coast of Monterey, California.*

Right: Dave Partlow proudly displays the company motto during the Noble Energy job.

"Global Diving: Keeping Divers Off the Street Since 1979."

GLOBAL DIVING COMPANY MOTTO

From left, Kyle Watson, Jeff Eyler, Kurt VanCampen, and Chris Stokes salvaging a tugboat on the Duwamish River.

and flip the barge over, but Global and Fred Devine ended up losing the bid. Still, the intertwined events served as obvious indicators of the quantum leap the company had taken in just the previous decade.

"It was interesting thinking back to those days and what a difficult project that seemed to us at that time," he said. "Now, we'd have bid that ourselves without thinking twice about it. The work was well within our capabilities."

The barge salvage was yet another example of the circular nature of the industry—that past jobs, opportunities, people, even offices had a tendency to come back in unexpected ways, often serving as a reminder of just how far Global and its divers had come.

Combining Skillsets on the *Catala*

Thanks to stricter regulations, better safety measures, and a stronger environmental conscience, there were few new oil spills for Global Diving to clean up in the late 2000s. However, they were called upon for plenty of oil response work, including one high-profile endeavor that would combine Global's distinct skillsets in an as-yet-untested way.

In 2006, Global began work removing oil from a vessel that sank with nine tanks full of fuel—and completed the project without a single dive, blending traditional diving salvage techniques with non-diving environmental response. That April a beachcomber in Ocean Shores, Washington, a popular vacation spot and habitat for endangered nesting birds, stopped along his way at the graffiti-covered wreck of the S.S. *Catala*. The 229-foot steamer, which at various points in its history was a transoceanic passenger boat, floating hotel, and casino, had run aground and tipped over during a storm 41 years earlier. There it had remained, largely buried in the sand at Damon Point State Park, trading hands between various salvors who never got her out. Poking a stick into an opening on the protruding hull's side of the dubious Ocean Shores landmark, the beachcomber probably thought nothing of it until the stick came back out covered in thick, black gunk.

The discovery of a potential oil spill sent the Department of

Setting Safety Standards Across the Industry

GLOBAL'S EFFORTS TO IMPROVE work-site safety and deliver higher-quality end products extend outside the company, and into industry organizations in which Global members take active leadership roles.

Global Diving & Salvage is an accredited member of industry-wide organizations across its service lines, including the American Salvage Association (ASA), the Spill Control Association of America (SCAA), the Association of General Contractors of America (AGC), and the Association of Diving Contractors International (ADCI). Devon Grennan is on the board of directors for SCAA, David DeVilbiss serves on ASA's executive board, and Bruce Humberstone is a member of ADCI's board of directors and Saturation Dive Safety Committee.

Tim Beaver served over 10 years on the board of directors of the ADCI, and from 2010 to 2012 represented Global on the select ADCI executive committee that overhauled the internationally recognized *Consensus Standards for Commercial Diving*. Beaver worked especially hard to make certain there was a balance between the needs of the offshore diving contractors and the inshore diving contractors.

Beaver served as president of ASA from 2011 to 2013, heading the country's premier group of commercial salvage companies, which banded together in 2000 to tackle salvage-related issues that impact the nation's maritime community.

Ecology and four other agencies scrambling to respond. They needed salvors who could dig in to get the *Catala*'s tanks, and then get rid of the oil, and they needed them there yesterday. Global fit the bill.

After an engineering subcontractor designed a 3-D model of the mostly buried vessel, work began excavating the *Catala* to get at her oil tanks. As backhoes removed sediment, significant amounts of materials containing asbestos were discovered and dealt with as they were revealed in the sand. Once the tanks were accessed, Global personnel used a boiler system to heat the "peanut butter" below, and then the warmed and less viscous (and more cooperative) oil was pumped out to vacuum trucks and transferred to larger tanks. Over 34,000 gallons of oil were removed during the project, which was completed in August 2007. Also removed was the hull of the *Catala*, finally raised from its longtime resting place.

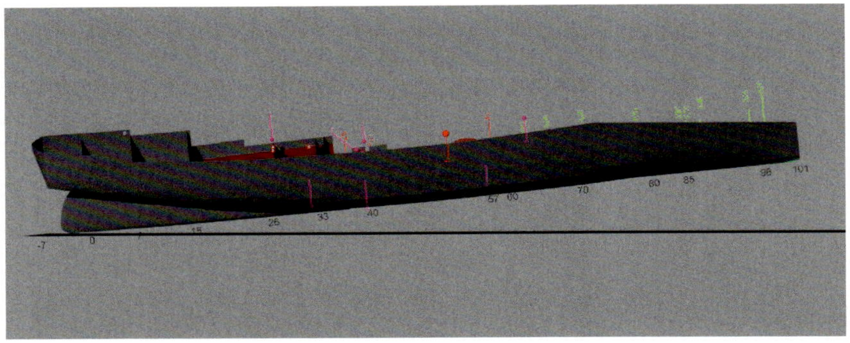

Top: The Catala *became a local attraction and preferred graffiti target after its grounding, as shown in this 1978 photo.*

Bottom: 3-D modeling was used to determine how far the vessel was buried under the sand, and at what angle.

Opposite: Global crews work on the excavation and removal of the Catala *during the company's first major job that involved no traditional dive work. Left to right: Jeff Eyler, Kelly Kriska, Ian Costin, Bill Akan, Jim Givan, and Nate McLellan.*

"The *Catala* is the shining example of what happens if you do what you say you're going to do, keep your nose clean, and know how to organize jobs," Tim Beaver said. "That one came to us and not to anybody else because we had a unique set of skills. It was a high-profile, high-risk job for the agencies, who needed to get it done right the first time."

The success of the *Catala* and subsequent projects combining traditional diving and marine environmental work showcased Global's unique ability to bring together diverse skillsets under one company umbrella. "Those skillsets—responding to oil spills, preventative booming, and derelict vessel removal—are intricately linked and they complement each other," Devon Grennan said. "They add nicely to our already strong foundation of commercial diving and construction experience, allowing us to provide an all-inclusive service offering to our clients."

Together
Everyone
Achieves
More

Old Faces in New Roles

Now that the company was recognized worldwide as an elite diving, salvage, offshore services, subsea construction, and environmental response operation, the divers who had been there since the days of root beer floats on Harbor Island were transitioning out of the water and into the office to support a rapidly expanding family at Global Diving & Salvage.

In 2009, Devon Grennan became president of Global Diving & Salvage, assuming the role long held by Tim Beaver. Grennan had been general manager since 2005. Beaver became chief executive officer, and company cofounder John Graham assumed his role as chief strategic officer.

For Beaver and Graham, it was time to loosen their grip on the reins of the company they had grown from two hoses to four saturation diving systems. Graham knew his company would be in good hands, not just with Grennan, but also his other hires. "I've always tried to hire people smarter than myself," he said. "A bunch of companies don't like smart people; they hire drones. I don't know how you survive that way."

Global personnel working on the Deepwater Horizon cleanup display a company mantra on their backs: "TEAM – Together Everyone Achieves More."

If the smart people he and Beaver hired didn't live up to company values or expectations, Graham felt he had no one to blame but himself. "Tim and I established, nurtured, and mentored these folks who are running with the scissors now, and we hope they're doing the right thing. If they're not, the way I look at it, it falls back on me. I'm the one who set them up and handed them the scissors."

After becoming president, Grennan worked to uphold the founders' philosophy regarding hiring and "handing off the scissors." "I'm not delegating because I don't want to do the work; I'm delegating because I've got a lot of people a lot smarter than I am working for me," he explained. "Because they're better at whatever it is they do. They understand the rough edges better than I do. I think we're perpetuating that as a company, and that's how the company grows."

To help continue the Global Way and manage a growing workload across a variety of project types, the company divvied projects up amongst service lines, each headed by a longtime Global diver promoted to a vice president position. Bruce Humberstone oversaw Offshore Operations, managing Global's many projects for the oil and gas industry; David DeVilbiss managed Marine Casualty and Emergency Response Services, furthering his work in marine salvage; and Mike Langen continued to "build better mousetraps" as head of Marine Construction, Engineering, and Technology. Additionally, Dan Pierson retired from the water to become VP of the Pacific Northwest region, in part, he thinks, "because I missed a meeting and got stuck with it."

DeVilbiss, Humberstone, and Pierson now took on a new challenge: managing divers, that unruly but talented breed. "The kind of ego that makes a good diver a great diver is the kind of demeanor that can be difficult to work with," DeVilbiss said. "Managing divers is much more difficult than diving ever was. But it's also challenging, and very rewarding. John Graham once told me that dealing with the personal side of things and the different personalities was the best part of the job, and the worst part of the job. I would have to agree with him after being on the management side of things for a while."

For Humberstone and Pierson, the greatest adjustment was spending more time out of the field and in the office. "I dove for 34 years, where I got in the water damn near every day, and I don't miss that one bit. But I do miss getting out in the field more," Humberstone said.

"I miss the diving, especially the camaraderie with the guys," Pierson said. "And in the office, you feel like nothing is ever completed. There's no single task that I ever feel like is put to bed."

Growing Beyond the Pacific Northwest

Humberstone, DeVilbiss, and Langen all would have their opportunities to get hands-on in the field, as well as with the always unfinished task of spreading the Global Way, through their work with Global's regional offices in Alaska, California, and Texas. Global Diving grew significantly in the 2000s, firmly establishing itself outside the Pacific Northwest with the addition of these three regional offices that support and work with Global's service lines on both regional and company-wide projects.

Global's office in Rio Vista, California, was its first expansion out of the Northwest, after the acquisition of Inshore Divers in 2004. Global won a large marine construction dive support contract working on the San Francisco-Oakland Bay Bridge and needed to staff the project. Tim Beaver and John Graham knew Inshore Divers owner Fred Aichele, an ex-Navy diver who was highly regarded in the industry but was ready to retire. In Aichele's company Global found a like-minded group who did high-quality, safe work and took care of their clients, and so acquired Inshore.

Once the project was over, the Global management up in Seattle was unsure how to proceed with their California counterparts. "We were naïve. We thought, We've got good people down there, so good luck!" Devon Grennan said. "From their perspective they thought, What the hell do you guys want us to do?

"We've learned that we have to take a much more active role, a management approach that is more collaborative. When that happens, everyone benefits because there are more opportunities for the company and the people there. It is an office that is now thriving and very busy." A second California office was opened in nearby Richmond in 2010, to act as a sister office to Rio Vista, focusing on environmental work. In 2012 the two offices became one to combine the diving services with the environmental. Global's California office is now consolidated on Mare Island in Vallejo.

Global bought Offshore Divers of Alaska in 2009 to open the Anchorage office. Alaska was a market that Global had wanted to enter for some time, and when a real estate broker representing Offshore called out of the blue regarding the potential sale

Devon Grennan thinks the mentality of a diver is similar to that of a race car driver, because of the tolerance for risk and thrill-seeker mindset necessary in both endeavors, and so it is no surprise that Global Diving has been involved in auto racing. The company, with John Graham leading the charge, sponsored the No. 41 car in the Grand-Am Rolex Sports Car Series. Global was one of three sponsors for the car along with Team Seattle and Dempsey Racing, headed by actor Patrick Dempsey, who also was one of the team's drivers. The racing team raised money through Team Seattle on behalf of Children's Hospital.

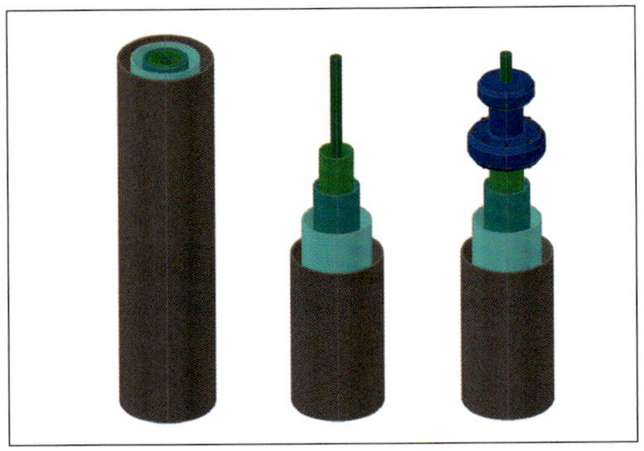

Top: Drawing with three images of a well head. Left to right: well casing cross section, conductors cut in a "wedding cake" formation, new well cap installed.

Middle: Conductors prepared for capping.

Bottom: The well cap in place.

of the business, Global jumped at the chance. After learning from its first acquisition, the company now recognized the importance of integrating the new regional offices into Global Diving's strong company culture. David DeVilbiss, an Alaska native, was given the task of managing the transition in Anchorage after Global's purchase of Offshore Divers.

"It was a real challenge on some levels," he said. "We spent a lot of resources bringing them down to Seattle to introduce them to everybody and see how we did our business. We tried to integrate crews as much as we could. That all happened at a period of the company where we were also trying to figure out how having regions and service lines was going to work ourselves. I think we've got a good product up there that we're proud to brand our own."

The office in Houston, Texas, opened in 2010. Global's workload in the Gulf increased once again due to the Bureau of Safety and Environmental Enforcement's Idle Iron policy, which mandated that offshore operators clean up and dismantle drilling platforms once wells were killed. Global divers were in high demand thanks to their proficiency in "wedding cake" well capping, a method of segmented drilling and intricate burning used to decommission oil wells (see diagram at left). The technique was learned and then mastered during the post-Hurricane Katrina cleanup. The amount of work in the Gulf led to Global building Sat IV—a fourth saturation diving system for a company that had only begun saturation diving a little over a decade earlier.

Global had looked at purchasing an existing company in the Gulf, but decided to instead open their own branch in Houston. Bruce Humberstone, a Gulf of Mexico veteran, led efforts to get the office going. "It's a slow process to build up your group, but we've got a reputation down there of being a very good, specialized company," he said. Humberstone also sees Global as having a distinct advantage over other companies in the Gulf that struggled during the economic recession of the late 2000s. "We're lucky in that if things are slow in the Gulf, other work in other regions can help maintain the company, which a lot of the oilfield diving companies don't have."

"I love the regional offices because we have these new people that are involved, with new ideas and outside experiences to complement everything we're doing as a company," Devon Grennan said.

Keeping Divers Diving

Global Diving needed the new people outside the Northwest to support a family of at times 350-plus, all with the same goal in mind: to keep the divers working, safely, and to give them the tools they would need to provide clients around the world with a quality end product.

As Global has increased its foothold in the niche of inland construction for dam projects, so too has grown the behind-the-scenes workload of a support staff dedicated to producing in-depth proposals for such work. Getting a proposal out the door is a time-intensive effort involving teams across Global's departments. Office staff develop hefty binders detailing why Global is capable of executing a project and how they will execute it. A proposal is part resumé, showing off past projects that highlight the company's experience in a given kind of work; the other part is the "how"—safety plans; project schedules; who will run the project and who will be in the water, plus their credentials; which subcontractors will be involved; even how sat equipment will be arranged on barges. Then comes the bid, based on man-hours, equipment rental costs, subcontractor rates, and any other costs—down to the exact cent.

The Cheesman Dam project in 2010 represented a significant achievement for support staff and the construction crew. Global's work on past inland construction projects, starting with Dworshak Dam and continuing with Gross Dam, provided strong portfolio pieces for the staff putting in long hours on the bid binder. Again working for Denver Water, Global crews assisted on an $18 million modernization project (the first since Cheesman was built 105 years previously) for the dam, which served nearly half of the 1.3 million Colorado residents in the greater Denver area. As soon as enough ice melted in the reservoirs in April, Global subcontractors began work drilling through the dam face and installing control piping. After 328

John Graham and Tim Beaver receive the Ernst & Young Entrepreneur of the Year award in the Transportation and Logistics category, 2011.

Above: Sat III worked at its highest elevation yet—6,842 feet—while at Colorado's Cheesman Dam.

Opposite: The project required both saturation and surface diving spreads at Cheesman Dam.

surface dives from April to July, saturation divers worked 24 hours a day, seven days a week until November to complete the replacement installation of three hydraulically operated gate valves on the dam's face.

The crew had to blast tunnels to enlarge them enough to accept three stainless steel spool pieces, one weighing 29,000 pounds and the other two 14,000 each. Jeff Martin, design project manager for Denver Water, issued a notice of completion in which he, on behalf of Denver Water, commended Global Diving & Salvage on a well-managed diving construction project with exceptional attention to detail and quality.

Cheesman solidified once again Global's sterling reputation in inland construction. With another notch in its belt, Global was able in 2013 to submit a proposal to Pacific Gas and Electric (PG&E) for a series of four dam projects in California. The proposal required overall safety and work plans, plus individualized plans for each dam—essentially five proposals in one. A team of 12 worked weekends and overtime to estimate costs, format documents, and make countless coffee runs to complete the dictionary-sized proposal securing the bid and the work.

Spills and Salvage

Global continued to respond to traditional spill and salvage work through personnel attending to emergencies for the Marine Spill Response Corporation (see sidebar, page 58) and through government projects targeting problematic shipwrecks. The Wreck Oil Removal Program (WORP), a federal program targeting sunken vessels with a high risk of discharging oil, has kept Global steadily involved in the removal of oil long held by the sea. WORP estimated at its outset in 2009 that there were some 7,000 wrecks around the coastal United States that were holding the equivalent of 15 *Exxon Valdez*-size spills

in total. Global has helped on several WORP responses, including the *Luckenbach* and the *Princess Kathleen* in 2010. The historic wreck of the *Princess Kathleen*, which ran aground north of Juneau, Alaska, in 1952, was the culprit in numerous oil leaks over the following decades, until Global divers removed over 140,000 gallons of oil from 14 tanks. Another WORP response, on the SS *Montebello* in 2011, required the use of remotely operated vehicles (see sidebar).

Global divers also traveled 5,200 miles west to the Philippines in 2010, to assist Titan Salvage in removing pesticides and other toxic chemicals from the tragic wreck of the *Princess of the Stars*. The passenger ferry ran aground and capsized during a typhoon, in one of the worst maritime accidents ever in the country. Several years later, Global personnel went east to Italy to help with recovery efforts for the *Costa Concordia*, the cruise ship that infamously ran aground in 2012 and lay on her starboard side in shallow Mediterranean waters. Global completed over 1,800 dives on the salvage and was the last diving subcontractor to leave the project before the wreck was secured for the winter of 2013 and then successfully moved upright nine months later, in September of that year.

Top: The defueling operation on the Princess Kathleen, *Lena Point, Juneau, Alaska.*

Bottom: ROV with sampling tool skid for the Montebello *assessment.*

Still Doing the Impossible

Over 35 years, Global Diving has been proud to tackle projects that others have thought impossible—sometimes including the clients for said impossible projects. Mike Langen recalled accepting an award on behalf of Global for a dam construction project. "The guy was shaking my hand and he said, 'Yeah, we had an office pool betting that you could never get this done!'"

Global is more proud of their commitment to their employees through tough times for both employee and company alike, a continuation of the company's pledge from its inception to not let its people "go backwards."

The economic recession of the late 2000s hit Global slightly later than other businesses. The company—whose workload became more and more regulatory driven—found that public entities had less

The ROV is lowered over the side of the Nanuq *to inspect the* Montebello*, some 900 feet below.*

Remotely Operated Oil Seekers

SINCE ITS DAYS ON HARBOR ISLAND, Global Diving has tinkered with remotely operated vehicles (ROVs) on projects including salvages, or just scouting out a site before divers get in the water. As camera and robotics technologies have increasingly improved in the last decade, ROVs and their small operational crews have found steady work on important projects, particularly in locating and removing oil on vessels resting deep on the ocean floor.

The 2009 Robson Bight salvage was Global's first partnership with Mammoet Salvage of Holland and the company's first chance to use the Cougar, a state-of-the-art ROV. A diesel fuel tanker sank to the bottom of the small bay on the northeast side of Canada's Vancouver Island; Robson Bight is also a protected killer whale habitat, and so salvage efforts were even more sensitive. The Cougar was lowered some 1,165 feet to where the tanker had lain for two years to carefully recover the tanker while containing fuel and oil. Mammoet and Global successfully completed the salvage without incident and with no additional environmental impact.

As part of a federal Wreck Oil Removal Program (WORP) project, the Cougar dove deep again in 2011 to search for oil on the SS *Montebello*, an oil tanker sunk by a Japanese torpedo in 1941 and thought to have gone to the bottom with 3 million gallons of oil aboard. The *Montebello* lay 900 feet underwater, 6.3 miles off the California coast and just south of the Monterey Bay National Marine Sanctuary. Using Global's patented remote drill sample and plug system, the Cougar found no evidence of oil on the ocean floor nor in the tanker, whose hulls were found to be structurally sound; researchers believed the oil dissipated over time and that some was released initially, but not enough to reach shore. The Cougar's findings were a huge relief to the many organizations that comprised the *Montebello* task force, including the U.S. Coast Guard, the National Oceanographic and Atmospheric Administration (NOAA), the California Department of Fish and Wildlife's Office of Spill Prevention and Response, and the Monterey Bay Aquarium Research Institute.

Giving Back and Helping Out

GLOBAL DIVING IS A COMPANY THAT SOLVES PROBLEMS. While most of its problem-solving efforts take place hundreds of feet underwater, every September since 2006, Global personnel and company partners have hit the links to help causes near and dear to the company's heart.

Global Diving's annual golf tournament has raised money for organizations including the Puget Sound Maritime Historical Society, dedicated to preserving and creating an appreciation for the region's maritime past, and Seattle Children's Hospital, whose mission is "to prevent, treat, and eliminate pediatric disease." Global's tournament raises upward of $50,000 a year, and proceeds from the event have been used to raise funds for a permanent maritime exhibit at Seattle's Museum of History and Industry; a blood draw clinic at Children's Hospital's Autism Center; and a Hybrid Surgical Unit for the Hospital's Heart Center. As of 2013, Global's charitable donations have totaled over $250,000.

Global Diving is also actively involved in the Seattle Propeller Club, an organization promoting marine commerce in the Puget Sound region. Additionally, the company has lent the *Prudhoe Bay* to even rowdier deckhands than usual for Seattle Seafair's annual pirate landing at Alki Beach.

"We participate in the maritime industry and the region because our people genuinely want to be involved," Devon Grennan said. "We want to be part of the maritime community."

The Global Diving team's dedication to their communities is not limited to planned events. In March 2006, Chris Butler, Ryan Cummings, and Travis Havens were diving on a job at the Port of Seattle's Terminal 18 when a nearby lifeboat capsized during a safety exercise. They immediately responded to the crisis, helping until the Coast Guard could arrive and saving four lives in the process. The three heroes were awarded special certificates of appreciation from the Seattle Fire Department during the department's annual awards night. "If you can help somebody, and you're there, you just do what you've got to do," Travis Havens told the *Seattle Post-Intelligencer*.

Top: Left to right, Clarence Banks, David Parr, John Graham and from Children's Hospital, Barbara Koler, Barbara Mann, Aileen Kelly.

Bottom left: From left, Ryan Cummings, Chris Butler, and Travis Havens receive a certificate of appreciation during the Seattle Fire Department's annual awards night.

Bottom right: Fundraising for Children's Hospital.

money for projects. "We knew it was inevitable that we were going to get hit by the recession, but we didn't feel it until our backlog of jobs had been completed," Trinity Ng-Yeung said.

The company's founders and management made tough decisions to be able to keep their team together and support their Global family, especially when dive work was the last thing on certain employees' minds. David DeVilbiss experienced this firsthand when he was diagnosed with Hodgkin's lymphoma in 2000. Global kept him on board throughout his treatment, including chemotherapy. He went to Tim Beaver and John Graham to tell them the news. "The first thing they said was, 'Don't go looking for a job. We've been needing an equipment manager. We're going to give you this much money a week; come in if you feel up to it.' They had more expectations than that, obviously, but they wanted me to still be a part of the company and were willing to work around my treatment. That's pretty cool."

A Global crewmember looks on at the shipwrecked Costa Concordia. *Global performed just under 1,900 dives, providing a 12-man crew on site for over a year before the cruise ship was parbuckled in September 2013.*

DeVilbiss has not been the only one to go through some tough times. Sales Manager J.P. Hallmark was diagnosed with non-Hodgkin lymphoma and underwent treatment for six months. The company kept his position open because, according to Devon Grennan, "it was the right thing to do. Many other large companies would've said, We can't do this. But we're not the other guys—we're Global. There's no consensus annual meeting where the shareholders get together and yell at each other about the share price dropping. You make decisions here that are for the benefit of everyone in the company, not a shareholder."

Salvage has undergone tremendous change, from surviving hand to mouth futilely patching sinking drydocks to completing multimillion-dollar government projects as a prime contractor. None of it would have been possible without dedicated personnel committed to safety, quality of service, and growth since the company's inception.

When the Skagit River Bridge, a heavily trafficked portion of Interstate 5 in Washington, collapsed after a truck struck a support structure, Global divers worked 24 hours a day for 13 days providing diving and salvage support under Atkinson Construction.

"In the early days, the founding members of the company were it," Jackie Lewis said. "They've all missed holidays, kids' birthdays, and school events because they were striving to succeed—and they did."

The spirit continues as of the writing of this book. Every year the company gives the Yoke Award—an actual goat yoke purchased from Arkansas on eBay—to employees who demonstrate a willingness to buckle down when they find themselves in the crosshairs of an exceptionally heavy workload.

"This is not a job that you kind of like," said Devon Grennan, who became CEO of Global at the end of 2013. Tim Beaver and John Graham stayed on as the company's founders. "If you don't love this, if you don't really take a lot of pride in the highs and recognize that there are going to be some real frustrations and stresses on your family, it will chew you up. This industry never turns off."

Though the industry is constantly changing and competition is always fierce, Grennan is adamant: "We know that if we maintain the safety of our operations, set a high standard in client satisfaction and quality services, and couple that with a clear identity of who we are, where we have come from, and what we can be, there will be tremendous opportunities for us as a company and for our employees."

Grennan expects there to be many such breaks in the future—as long as Global continues to have people who can recognize those chances and have the willingness to seize them, as risky as they might be. "It is an exciting time to be at Global right now," he said. "There are a lot of opportunities on the horizon, and we are transitioning ourselves from a traditional commercial diving company into a more well-rounded marine and subsea service provider. There is still the core belief that a scrappy group of divers got together and built something that was not initially expected of them, that was a little different. We have people now who still believe in that approach, and we are adding new individuals every day with that same mentality. We have a great reputation, great people, and as far as I am concerned, the sky is the limit."

The company has been fortunate to bring together a crew of individuals who are passionate about the diving industry. The family, and the scope of the company's work, has grown larger than the founders could ever have imagined.

"It's a hell of a story," Tim Beaver said. "I never dreamed that we'd be able to do what we do today."

The Global Way established by the company's founders drives the company, even as Beaver, Graham, Lewis, and others from Global's beginnings grow into new roles.

"It's not that different from the first time I saw a crew drive out in a truck, without me with them," John Graham said jokingly. "I thought, Jeez, I hope they bring my damn truck back in one piece."

"Presented annually to the employee who best exemplifies the Global spirit by their willingness to 'shoulder the load.'"

Yoke Award Recipients:

2006: Sarah Burroughs

2007: Dan Pierson

2008: Trinity Ng-Yeung

2009: Matt Nienow

2010: Shawntel Johnson

2011: Dean Stone

2012: Brooke Boge

Personnel on Major Projects

Brother Jonathan

(1997, 2000)

Round One, 1997

Doug Alies
Joe Amaral
Tim Beaver
David DeVilbiss
Scott Dimmick
John Graham
Bruce Humberstone
Ronald Larsen
Don Moore
Dave Partlow
Daniel Pierson
Herb Williams

Round Two, 2000

Tim Beaver
Wade Bliss
Mike Dennis
Chris Drakos
Jimmy Geissler
Terry Geissler
John Graham
Travis Havens
Bruce Humberstone
Jon Jeffrey
Ronald Larsen
Marc Mittry
Dave Partlow
Daniel Pierson
Dennis Tsugawa
Mike Wheeler

New Carissa

(1999)

Tim Beaver
Glen Beckman
Thom Davis
David DeVilbiss
Jeffrey Eyler
Skip Green
Devon Grennan
Ryan Grennan
John Hallmark
Ronald Larsen
Paul Leanderson
Jackie Lewis
Adrian Lipp
Brent Lyngstad
Daniel McBride
Nathan Nelson
Daniel Pierson
Robert Seasock
Wayne Tayler
Kurt VanCampen

Luckenbach

(2002)

Tim Beaver
Wade Bliss
Patrick Boyd
Joseph Cadwell
Ryan Cummings
David DeVilbiss
Chris Drakos
Nicholas Finney
David Gill
Travis Havens
Donald Hosford
Jon Jeffrey
Ronald Larsen
Michael McNeil
Dave Partlow
Daniel Pierson
Michael Shea
Dennis Tsugawa

Ensco 64

(2005)

Steve Ackart
Nyibal Aj Adams
William Akan
Anthony Anderson
Collin Anstrom
Mark Bailey
Matthew Ballard
Duane Black
Dennis Blank
Patrick Boyd
Rodrigo Boyle
Grant Bradley
Shane Brock
Curtis Bulka
Randy Busby
Joseph Carp
Jarrod Christensen
Timothy Christiansen
Eric Cousin
Jeff Cox
Thomas Crites
Aaron Dack
Spencer Dell
David Dettmann
Kevin Doern
Landon Erickson
Kenneth Faulkner
Nicholas Finney
Eugene Flennaugh
Hiram Garcia
James Geissler
Terry Geissler
David Gill
Paul Guiou
James Hayes
Darryl Heath
James Hegeman
Derek Herness
Mark Hintz
Michael Hofmann
Kristofer Homeier
Donald Hosford
Brian Huber
Bruce Humberstone
Samuel Humphrey
Brian Jamison
Jimmy Johnson Jr.
John Lapeyrouse Jr.
Robert Lapeyrouse
Ronald Larsen
Jules Lemaitre
Mathew Lohff
Patrick Lombardi
Dale Look
Steve Majorie
Alejandro Mandujano
Andrew Maul
Edgar McAfee
Scott McAuley
Michael McCrory
Conner McLeod
Paul Melby
Jason Mitchell
Marc Mittry
Kyle Nordgren

Nicholas O'Brien
Mark Oglevie
Mark Olson
Jack Opie
Jose Orengo
John Ortiz
David Partlow
Adam Petra
Joseph Pick
Daniel Pierson
Richard Rivers III
Peter Roulet
John Sabo
Richard Satalich
George Saunders
Don Schuyler
Jesse Scott-Kandoll
Riley Smith
Reed St. John
Shawn Stambaugh
Bryan Sully
Ronald Taylor
Stephen Testroet
Dennis Tsugawa
Chad White
Herb Williams
Wayne Williamson
Slawomir Wozniak
Russell Wright

Catala

(2006-2007)

Rich Brandt
Christopher Butler
Ernest Cheeka III
Ian Costin
David DeVilbiss
Jeffrey Eyler
Jeff Frye
Michael Garza
James Givan
Colins Goertzen
Devon Grennan
Jack Herron
Rick Klebanow
Kelly Kriska
Ronald Larsen
Mac McCarthy
Nathan McLellan
Kyle Nordgren
Chris Stokes
Guadalupe Vargas Jr.
Joshua Walker

Shaft 6

(2007-2012)

William Akan
Matt Ballard
Bob Beamis
Eric Bittick
Robert Bouis Jr.
Patrick Boyd
Christopher Butler
Tiffany Cartier
Dean Catalano
Shane Coffelt
Zach Coots
Thomas Crites
Ryan Cummings
Carl Damgaard
Rick Davis
Allen Deaver
Spencer Dell
Kevin Doern
Chris Drakos
Brady Dugas
David Dukes
David Edmisten
Landon Erickson
Nicholas Finney
Ryan Fitzsimmons
Gary Galstad
Abraham Rodriguez Garcia
James Geissler
Terry Geissler
Daniel Gilchrist
David Gill
Paul Guiou
Chris Hackworthy
Travis Havens
Bryan Hayden
James Hayes
Darryl Heath
James Hegeman
Raymond Heroy
Mark Hintz
Michael Hofmann
Cameron Holm
Kristofer Homeier
Brent Hookom
Donald Hosford
Peter Jobes
John Lapeyrouse Jr.
Robert Lapeyrouse
Michael Lucia
Carl Lundblade
James Lyons
Steve Majorie
Kirk Marshall
Edgar McAfee
Peter McCracken
Marc Mittry
Michael Moschetto
David Murphy
Matthew Nienow
Mark Oglevie
Jose Orengo
David Partlow
Bruce Pastre
Joseph Pick
Daniel Pierson
Jeff Pietras
Timothy Pinczes
Alex Ramsey
Richard Rivers III
Travis Rudolph
John Sabo
Joseph Sabo
Richard Satalich
Ryan Smith
James Stock
Dean Stone
Dennis Tsugawa
Matthew Vancleave
Jeffrey Wanner
Wayne Williamson
David Zenicki

Index

Page numbers in *italics* indicate photos.

Photo Credits

All images courtesy of the Global Diving archive except as follows:

7: Tim Beaver
14a: Science Museum/Science & Society Picture Library
15: OAR/National Undersea Research Program
16a: Museum of History & Industry
16b: Courtesy Sarah Burroughs
18a: Katie Christensen, katiekapow.com
18b: Greg Harem
19: Greg Harem
21: Puget Sound Maritime Historical Society
25a: Thom Davis
25b: Greg Harem
28: Greg Harem
31: *Post-Intelligencer* Collection, Museum of History & Industry
35b-c: Sea Form Manufacturing Inc.
36a: Greg Harem
42: Greg Harem
43: Greg Harem
48: Joel W. Rogers, joelrogers.com
49b: Greg Harem
51b: Gary Duff
53a: Thom Davis
55a: Greg Harem
56: Natalie Fobes
61: Courtesy Marine Pollution Control
64: Carl Cook
65a: Greg Harem
67: Mac McCarthy
70c: U.S. Army Corps of Engineers
73: Puget Sound Maritime Historical Society
76a: San Francisco Maritime National Historical Park
76b: Del Norte County Historical Society Photograph
77b: David Lawrence Rare Coins
83: Auctions by Bowers and Merena, Inc.
84: Tim Beaver
86-87: Tim Beaver
88: Environmental Protection Agency
89: University of Washington Libraries, Special Collections, UW36132
90: Devon Grennan
93b: Kristian Sanford
94: Jelson25
95a: Devon Grennan
95b: Devon Grennan
96a: Devon Grennan
96b: Devon Grennan
97: Devon Grennan
98: Devon Grennan
102a: Courtesy Chelan County PUD
104a: Devon Grennan
104b: Tim Beaver
109: Denver Water
114a: *Rigzone* Magazine
114b: National Oceanic and Atmospheric Administration
120a: Bureau of Ocean Energy Management
124: New York City Department of Environmental Protection
125: New York City Department of Environmental Protection
138a: *Post-Intelligencer* Collection, Museum of History & Industry
139a: Washington State Department of Ecology, courtesy Global Diving
149: Robert Schwemmer, National Oceanic and Atmospheric Administration - National Marine Sanctuaries
159: Kristian Sanford

about the author

JOE GUSTAV is an author, writer, and native West Seattleite. His work has been featured in Seattle Met *Magazine. Joe received his degree in English from Whitman College in Walla Walla, Washington. He is a beginning sport diver, diehard Seattle sports fan, and mentor at the 826 Seattle writing center.*

THE SUCCESS OF A DIVE PROJECT may ultimately rest on the shoulders of the person in the water, but it would not be possible without an extensive support team. For me, the completion of this history was no different.

Thanks to Tim Beaver and John Graham for your devotion to and passion for this book. I hope it does justice to your stories, and to your company. Thanks also to Devon Grennan, for your input and good humor; and to Jackie Lewis, for keeping this unruly group of divers in line (and me too).

I'd especially like to recognize the contributions of Global's marketing team, Frank Immel and Jimena Castria, in providing direction, feedback, and assistance. Thank you for driving this book forward.

I am indebted to the many people interviewed for this book, who enthusiastically and graciously shared their time, memories, and photos to help tell Global's story.

My gratitude goes to Petyr Beck of Documentary Media for heading this project and including me in it every step of the way, and to Marilyn Esguerra of Philips Publishing, the designer of this handsome book.

breAnne and Silas at Robert Spector Consulting, I appreciate your support and encouragement in this project. And to Robert, the best mentor a young writer could ever ask for—thanks, coach. Preparation truly does create separation.

Mom, Dad, Andrew, Nana, and Auntie—thank you for being my loving and supportive family.

Finally, thanks to everyone in the Global Diving & Salvage family. I am honored to tell your story. Be safe, keep doing amazing work, and stay as dedicated to getting out of your comfort zones and attacking new opportunities in your next 35 years as you were in your first 35.

Joe Gustav